Lecture Notes on Mathematical Modelling in the Life Sciences

The rapid pace and development of new methods and techniques in mathematics and in biology and medicine creates a natural demand for up-to-date, readable, possibly short lecture notes covering the breadth and depth of mathematical modelling, mathematical analysis and numerical computations in the life sciences, at a high scientific level.

The volumes in this series are written in a style accessible to graduate students. Besides monographs, we envision the series to also provide an outlet for material less formally presented and more anticipatory of future needs due to novel and exciting biomedical applications and mathematical methodologies.

The topics in LMML range from the molecular level through the organismal to the population level, e.g. gene sequencing, protein dynamics, cell biology, developmental biology, genetic and neural networks, organogenesis, tissue mechanics, bioengineering and hemodynamics, infectious diseases, mathematical epidemiology and population dynamics.

Mathematical methods include dynamical systems, partial differential equations, optimal control, statistical mechanics and stochastics, numerical analysis, scientific computing and machine learning, combinatorics, algebra, topology and geometry, etc., which are indispensable for a deeper understanding of biological and medical problems.

Wherever feasible, numerical codes must be made accessible.

Founding Editors:

Michael C. Mackey, McGill University, Montreal, QC, Canada

Angela Stevens, University of Münster, Münster, Germany

More information about this series at http://www.springer.com/series/10049

Frederic Guichard · Justin Marleau

Meta-Ecosystem Dynamics

Understanding Ecosystems Through
the Transformation and Movement of Matter

 Springer

Frederic Guichard
Department of Biology
Faculty of Science
McGill University
Montreal, QC, Canada

Justin Marleau
Department of Biology
Faculty of Science
McGill University
Montreal, QC, Canada

ISSN 2193-4789   ISSN 2193-4797 (electronic)
Lecture Notes on Mathematical Modelling in the Life Sciences
ISBN 978-3-030-83453-1   ISBN 978-3-030-83454-8   (eBook)
https://doi.org/10.1007/978-3-030-83454-8

This Springer imprint is published by the registered company Springer Nature Switzerland AG
The registered company address is: Gewerbestrasse 11, 6330 Cham, Switzerland

Preface

Over the last 15 years, theories integrating ecosystem and spatial dynamics have quickly coalesced into meta-ecosystem theories. Ecosystem dynamics, going back to the work of Odum [135], refers to the cycling of matter and energy across ecological compartments through processes such as consumption and recycling. Spatial dynamics established its ecological roots with island biogeography and metapopulation theories, and focuses on scaling up local ecological processes through the limited movement of individuals and matter. This book presents current meta-ecosystem models and their derivation from classic ecosystem and metapopulation theories. We show how the interaction between the cycling and movement of matter is a ubiquitous property of natural ecosystems and has far reaching implications for the coexistence of species, ecosystem productivity, and their response to stresses and environmental change. Specifically, we review recent modelling efforts that have emphasized the role of nonlinear dynamics on spatial and food web networks, and cast their implications within the context of spatial synchrony and ecological stoichiometry. We finally suggest that these recent advances naturally lead to a generalization of meta-ecosystem theories to spatial fluxes of matter that have both trophic and non-trophic impact on species. Such integration brings together areas of research, such as behaviour, chemical ecology, food web theories, and landscape ecology, that have rarely overlapped in the past.

This book covers mathematical models that have contributed to meta-ecosystem theories in the ecological literature. We limit our discussion to work that has been labelled as meta-ecosystem theory while acknowledging the many theories and models that have addressed similar questions (ecological implications of cycling and movement of matter) without being labelled as such. While the concept of meta-ecosystems has emerged less than 20 years ago [109], spatial ecosystem dynamics has been studied as part of ecosystem ecology through flows of matter across macro-habitats and through large-scale quantitative ecosystem models [66]. There is also a rich body of literature studying models of pelagic ecosystem dynamics, and integrating both the recycling and movement of nutrients (NPZ models, [43, 91, 168]). Biogeochemistry has played a central and leading role in the study of elemental cycling and spatial fluxes within and across ecosystems [158]. Because of its origin in metapopulation theories, meta-ecosystem models can contribute ecological insights to the biogeochemistry perspective. For example, meta-ecosystem theories place a strong emphasis on feedbacks between animal population dynamics and fluxes of matter, an emphasis generally missing in biogeochemistry. We hope to outline a coherent and specific approach to ecosystem dynamics that can (1) provide unique insights by focusing on qualitative predictions that lend themselves to experimental and field validation across systems and (2) contribute to a more mechanistic understanding of ecosystem dynamics in general and to their response to human impact.

The meta-ecosystem models we cover in this book were formulated as extensions of metapopulation and metacommunity theories that are thus briefly reviewed as the basis for a minimal meta-ecosystem model. Much of this material was also covered by Loreau [106]. Using this minimal model as our backbone framework, we define meta-ecosystems models as those integrating spatial fluxes and cycling of inorganic (and organic) matter to the dynamics of living compartments such as primary producers and consumers. Our goal here is to illustrate how these flows of matter interact with mechanisms of population growth and species assembly that are core to metapopulation and metacommunity theories. It is by building on these core population and community theories that a meta-ecosystem framework can make its unique contributions to theoretical ecology. We refer the reader to other readings for a more in-depth treatment of metapopulation and metacommunity models. We also focus on a significant but particular subset of meta-ecosystems theory and limit our presentation to mathematical models of ecosystem dynamics. While we put a number of predictions in the context of natural systems, our coverage of meta-ecosystem dynamics leaves out part of meta-ecosystem theory, which includes work based on experimental and statistical approaches that are not reported here. A few recent reviews will provide readers with system or question focused overviews of experimental meta-ecosystem studies.

We believe a mechanistic and dynamic approach contributed by meta-ecosystem theory is filling an important gap in current research on the loss of biodiversity that has been associated with the degradation of ecosystem functions. While this problem receives much attention from scientists, policy makers, and from the public, the relationship between biodiversity and ecosystem function could benefit from strengthening its connection (1) with a mechanistic understanding of ecosystem dynamics, (2) with ecosystem functions that includes the cycling of matter, and (3) with theories that account for the strong variability displayed by many natural systems. One of our goals is to show how simple dynamic meta-ecosystem models can provide predictions based on feedbacks linking biodiversity and ecosystem functions rather than considering the degradation of ecosystem function as a consequence of biodiversity loss. We believe this more integrative perspective on the current biodiversity crisis can be most successful if we understand the integration between the cycling and movement of organic and inorganic matter that are core components of ecosystem functions.

In each chapter, we presents models, and start from a minimal model configuration that allows studying a property of interest (e.g. nonlinear functional responses). We

Fig. 1 Outline of our chapters and their interrelationships between different fields of ecology and applied mathematics

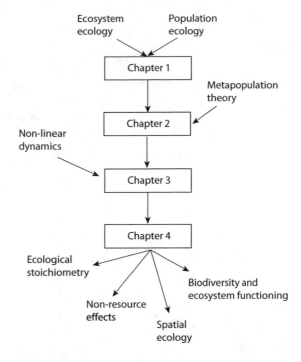

then present models studied in the recent literature followed by their analysis leading to specific ecological insights. More than resulting in a condensed summary of recent theoretical work, our goal is to identify commonalities and to outline a common theoretical framework that can help the future development of meta-ecosystem theories pursuing both generality and applicability. This book is intended for theoretical ecologists or for applied mathematicians interested in ecosystem dynamics. However, most of the text should be accessible to most ecologists, from advanced undergraduate students to more senior scientists.

Meta-ecosystem theory has emerged from the incorporation of insights from multiple fields of ecology and applied mathematics (Fig. 1). The first two chapters introduce the meta-ecosystem framework, first from the perspective of elemental cycling models (Chap. 1) and then from a spatial perspective, as a extension of metapopulation theory (Chap. 2). The remaining two chapters explore specific properties and predictions of meta-ecosystem models and are not intended to be read in a particular order. These properties include nonlinear and non-equilibrium dynamics with the analysis of spatial synchrony (Chap. 3), topological of spatial networks, multiple nutrient limitation and stoichiometry, and non-trophic interactions (Chap. 4).

Montreal, Canada   Frederic Guichard
December 2020   Justin Marleau

Acknowledgements

This book is the result of 15 years of collaborative work on meta-ecosystem dynamics, involving a number of researchers who have collectively given meta-ecosystem theory an identity and a place in ecological science. We were constantly stimulated and inspired by their creativity and insights and we are thankful for this opportunity to report their work. Their names appear abundantly throughout the book. Particular thanks to Isabelle Gounand and Shawn Leroux for their thoughtful comments on an earlier draft of this book. The authors also wish to acknowledge financial support from the Natural Science and Engineering Research Council (NSERC) of Canada through their Discovery program.

Contents

Symbols Used in Equations

Meta-ecosystem theory, due to its origins as a synthesis of ecosystem, metapopulation, metacommunity and landscape ecology, has been developed with numerous models that do not use notation consistently across papers. For example, D can mean a stock of detritus [54, 57] or it can mean a disturbed mussel bed [86]. Furthermore, meta-ecosystem models are typically parameter rich, as a number of ecosystem processes are modelled in detail.

To reduce reader confusion, we provide here a list of symbols, their definition(s), and where they are used in the text when there are multiple definitions. We recognize that this list is quite long and involved, but comprehensiveness is needed with such a diversity of models. To simplify some notation, we use a general index of Z subscripts involving the identification of ecosystem compartments (i.e. I_Z refers to I_N, I_R or I_S, where N is a generic limiting available nutrient, while R and S are two potentially limiting nutrients). We also drop the other subscripts that refer to a given local ecosystem, such as i or x, as well as those that refer to certain generic identities of compartments, such as i, w, o and e, as these indexes are best explained by the surrounding text around the equations.

α	Influx of dead organic matter (Eq. 1.2)
α_Z	Maximum uptake rate for ecosystem compartment Z
β	Rate constant describing removal of dead organic matter (Eq. 1.2) or constant amount of sequestered matter (Eq. 4.13)
β_Z	Half-saturation constant for ecosystem compartment Z
χ	Population chemotatic sensitivity coefficient
χ_N	Proportion of nutrients recycled back into the available pool
δ	Spatial coupling strength
δ_Z	Mortality rate for biotic ecosystem compartment Z
Δ_Z	A generalized dispersal function for ecosystem compartment Z
ϵ	Consumption efficiency (Eq. 1.2) or non-dimensional time-scale parameter (Eq. 3.10)
ϵ_Z	Recycling coefficients for biotic ecosystem compartment Z (Eq. 4.1) or consumption efficiencies on primary producer Z (Eq. 4.8)
η	Non-dimensional predator mortality

γ	Periodic orbit of a patch (Eq. 3.12) and proportion of nutrients not assimilated by a consumer (Eq. 4.1) or proportion of nutrients assimilated (Eq. 4.11)
λ	Eigenvalue (Eqs. 3.7, 3.8, 4.5 or density-independent per capita growth rate (Eqs. 2.8 and 2.9)
ω	Constant matter leaking rate
Ω	Instantaneous amplitude of oscillations
ϕ	Phase deviation variable
Φ	Meta-ecosystem effect on density-independent growth rate
ψ	Death rate of a population
ρ	Birth rate of a population (1.1) or consumer nutrient quotient (Eq. 4.10) or supply of (in) organic matter (Eq. 4.12)
θ	Phase variable
Θ_Z	Amount of an element Z in consumer biomass
$\alpha\beta$	Maximum disturbance rates
$k\beta$	Minimum disturbance rates
a	Maximum prey consumption rate
a_{CZ}	Attack rate of consumer C on primary producer Z
A	Autotroph stock or biomass
b	Half-saturation constant (Eqs. 3.1 and 3.3) or colonization of individuals of a species (Eq. 2.2) or molar carbon of one individual consumer (Eq. 4.8)
B_Z	Biomass of ecosystem compartment Z
c	Consumption rate of raccoons (Eq. 1.2) or colonization rate from occupied patches (Eq. 2.1) or number of seeds produced per unit biomass (Eq. 2.16) or coefficients of the connectivity matrix (Eqs. 4.1 and 4.8)
c'	Colonization rate
C	Consumer abundance, biomass or nutrient stock
$\mathbf{C}$	Connectivity matrix
d_Z	Diffusion rate of ecosystem compartment Z
$\mathcal{D}$	Dilution rate
D	Detritus (dead organic matter; Eqs. 1.2, 2.7a and 2.14a) or disturbed mussel bed (Eq. 4.12)
D_Z	Diffusion rate for population Z
E	Amount of (in) organic matter sequestered in the mussel bed
E_Z	Output for ecosystem compartment Z
f	Local population dynamics function (Eq. 4.16) or recruitment of mussels (Eq. 4.14)
f_Z	Consumption function for biotic ecosystem compartment Z
F	Regional fluxes of matter
$\mathbf{F(x)}$	Vector-valued function describing local ecosystem dynamics across meta-ecosystem

g	Chemoattractant dynamics function (Eq. 4.16) or rate of disturbance propagation (Eq. 4.12)
G	Local cycling of matter (Eq. 2.4) or phase difference function (Eq. 3.16)
h	Non-dimensional prey abundance (Eqs. 3.9 and 3.10) or half-saturation level of matter density (Eq. 4.13)
h_{CZ}	Handling time of consumer C on primary producer Z
H	Herbivore or prey abundance (Eqs. 3.1 and 3.3) or 2π-periodic function (equations) or a general consumption function (Eq. 4.11)
I_Z	Nutrient input into a local ecosystem
j	Element of the Jacobian matrix
$\mathbf{J}$	Jacobian matrix
k	Non-dimensional carrying capacity
K	Carrying capacity
K_Z	Half-saturation constant for primary producer Z
l_Z	Loss rate for biotic ecosystem compartment Z
L_Z	Loss rate function for biotic ecosystem compartment Z
m	Extinction rate (Eqs. 2.1, 2.16 and 2.19a) or rate of loss of individual species (Eq. 2.2)
m_Z	Spatial flux rate for ecosystem compartment Z (Chap. 4) or loss rates for ecosystem compartment Z (Chaps. 2 and 3)
M	Mussel-covered area
$\mathbf{M}$	Spatial flux (or diffusion or movement) matrix
N	Available nutrient concentration
N_Z	Nutrient stock of ecosystem compartment Z (Eq. 4.1) or available nutrient Z concentration (Eq. 4.11)
p	Proportion of patches occupied by populations (Eq. 2.1) or non-dimensional predator abundance (Eq. 3.9) or larval recruitment of mussels (Eq. 4.12)
P	Population abundance of a species (Eq. 2.2) or primary producer (Eq. 2.7a)
$\mathcal{P}_Z$	Amount of element Z in primary producer biomass
q_Z	Fixed nutrient Z quota for a primary producer or consumer
Q_Z	Variable nutrient Z quota for a primary producer
r	Intrinsic growth rate of the population (Chap. 1 and Eq. 3.1), or mineralization rate (Eq. 2.7a)
r_Z	Maximum growth rate of primary producer Z
S	Constant total nutrients of an ecosystem 3.4 or total number of species in a metacommunity (Eq. 2.2) or constant nutrient supply rate (Eq. 4.8)
T_N	Variable total nutrients in an ecosystem
u	Population density in time and space
U	General nutrient uptake function (Eq. 4.11) or unoccupied mussel patch (Eq. 4.12)
U^D	Disturbed mussel bed (Eq. 4.12)
v	Chemoattractant concentration in time and space

V	Available habitat for species in metacommunity
$\mathbf{V}(k)$	Set of k matrices involving the spatial flux matrix and Jacobian matrix
W	Spatial coupling function
$x(t)$	Population size at time t
X	Size of an ecosystem compartment (Eq. 2.4) or a non-dimensional population abundance (Eqs. 3.10 and 3.11)

Chapter 1
Introduction: General Ecosystem Dynamics

Abstract Theories of population and community dynamics in ecology have been built on open systems where sources of resources and energy are external. In contrast, ecosystem theory studies the explicit contributions of external resources as well as internal sources coming from the local recycling of biomass. We start from the open logistic population to include 'ecosystem' dynamics models that can result in a closed system. The relative importance of local recycling compared to external flows of resources can have strong dynamical implications by imposing a negative feedback regulation through the conservation of mass within a closed ecosystem. From a minimal ecosystem model defined by Loreau [107], we review its integration to food web theories of species interactions developed by DeAngelis and Loreau [37, 107]. These models of local ecosystem dynamics have led to predictions linking the complexity of species interactions to ecosystem functions that derive from fluxes of matter across ecological compartments. Linking the diversity of species to functions at the ecosystem level has become a central goal of ecosystem and meta-ecosystem theories.

1.1 From Populations to Ecosystems

There is no 'right' way to look at an ecological system [28, 143, 149, 150]. Ecologists come from a variety of backgrounds and trainings, leading to a pluralistic field with diverse methodologies and theories concerning the same phenomena [143, 156]. The avalanche of theories, concepts and subfields can be daunting to any entrant to the field, perhaps leaving one feeling that 'anything goes' in ecology [28, 97, 119].

However, there are ways of organizing and reconciling the various approaches to ecology that can help us synthesize, integrate and potentially unify our explanations for why species X is found at density Y in location Z and/or how fast nutrient X cycles through ecosystem Y [143, 150, 156]. One useful organizing axis was first proposed by G. Evelyn Hutchinson, who suggested that ecological systems can be studied using a *biodemographic* or *biogeochemical* approach [80]. When we care about the number of individuals, be it from one, two or many populations of one, two or many species, then we are working within the biodemographic approach. When

© The Author(s), under exclusive license to Springer Nature Switzerland AG 2021
F. Guichard and J. Marleau, *Meta-Ecosystem Dynamics*, Lecture Notes on Mathematical Modelling in the Life Sciences, https://doi.org/10.1007/978-3-030-83454-8_1

we care about substances, say carbon, moving through the different components of the ecological system, then we are working within the biogeochemical approach. In modern ecological work, the biodemographic approach is generally labelled *population ecology* and the biogeochemical approach is called *ecosystem ecology*.

So, let us work through an example. Say we are interested in the population of raccoons in Mount Royal park (located in Montreal, Quebec, Canada) over the years. We let $x(t)$ be the density of raccoons at time t. The rate of change in the raccoon population, assuming no immigrant raccoons from outside the park and assuming no raccoons emigrate outside of the park, will be determined by the birth rate, P, and the death rate, Ψ. The birth and death rates are functions of density, i.e. $P = P(x(t))$ and $\Psi = \Psi(x(t))$, which we could determine empirically by tracking the population over time. Figure 1.1 shows simulated data showing a possible density relationship and total annual births and deaths (Fig. 1.1a). We can then fit functions to this data, and it turns out the relationships can be described as $P = \rho_1 x(t) - \rho_2 x(t)^2$ for births and $\Psi = \psi_1 x(t) + \psi_2 x(t)^2$ for deaths. Thus, if we examine the raccoon population's rate of change on an annual basis, we have the following equation:

$$\frac{dx(t)}{dt} = P - \Psi = \rho_1 x(t) - \rho_2 x(t)^2 - \psi_1 x(t) - \psi_2 x(t)^2 \tag{1.1}$$

This equation may look unfamiliar at first, but it can easily be rearranged into a well-known form in a few steps:

$$\frac{dx(t)}{dt} = \rho_1 x(t) - \rho_2 x(t)^2 - \psi_1 x(t) - \psi_2 x(t)^2$$
$$\Longleftrightarrow = x(t)(\rho_1 - \rho_2 x(t) - \psi_1 - \psi_2 x(t))$$
$$\Longleftrightarrow = x(t)(\rho_1 - \psi_1 - \rho_2 x(t) - \psi_2 x(t))$$
$$\Longrightarrow = x(t)(\rho_1 - \psi_1)\left(1 - \frac{(\rho_2 + \psi_2)x(t)}{\rho_1 - \psi_1}\right)$$
$$\Longrightarrow = rx(t)\left(1 - \frac{x(t)}{K}\right)$$

where r is equal to $\rho_1 - \psi_1$ and K is equal to $(\rho_1 - \psi_1)/(\rho_2 + \psi_2)$. The parameter r is commonly called the intrinsic growth rate of the population and represents the density-independent birth and death rates while the parameter K is called the carrying capacity and represents the density-independent birth and death rates scaled by the density-dependent rates. This formulation using r and K is called logistic growth and is used to describe the demography of ecological populations. One of the major advantages of this formulation is that it quickly allows the reader insight into the fundamental laws of population ecology [168]. When $x(t)$ is very small relative to K, then $(1 - x(t)/K) \approx 1$ and the rate of change is solely determined by $rx(t)$. Under these conditions, which assumes that individuals are exposed to identical and stationary environmental conditions, the population will grow (or decay) exponentially [168].

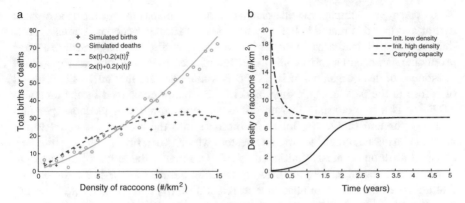

Fig. 1.1 Fictionalized population dynamics of raccoons on the Mount Royal, Montreal, Quebec, Canada. **a** Estimates of birth rates and death rates as a function of population density data. **b** Temporal population dynamics of the raccoons assuming low (solid black line) and high (dashed-dotted black line) initial population densities. Note that all solutions asymptotically converge to the carrying capacity (dashed red line)

However, environments rarely remain stationary for long and individuals can start experiencing deteriorating conditions as the population grow, such that $dx(t)/dt < 0$ for an $x(t) > x^*$ for all known populations, leading to the principle of self-limitation [168]. This principle is exemplified in logistic growth by the carrying capacity, which is the x^* for this model. Thus, depending on the initial population size, our modelled raccoon population experiences initial exponential growth (or decay) followed by a slow convergence towards the carrying capacity (Fig. 1.1b).

One of the limitations to a population ecology approach now comes into view: how can we determine r and K as environments change? Statistical fits to time series data and/or careful accounting of births and deaths across many populations can provide estimates for current and past r and K values, but these parameters are only partially related to the organism itself. The number of raccoons next year is not just a function of current raccoons, but also a function of the resources that are available to them in an area. If park attendants suddenly installed protective lids on all the garbage cans on Mount Royal, the amount of food available to the raccoons would suddenly plummet, leading to starvation (increasing deaths and reducing births). At this point, it can be helpful to reconceptualize our population of raccoons by using the concepts of ecosystem ecology.

Instead of viewing our raccoons as individuals that compose a population, we can view the raccoons collectively as a stock. The changes in the stock are due to fluxes into and out of the stock. What the stock is depends on the research question, but much of early ecosystem ecology focused on stocks and fluxes of energy. While direct measurement of energy in joules or calories is relatively rare, ecologists use the exchange and storage of carbon and/or biomass as a convenient proxy for energy. Therefore, for our raccoons and the refuse that they eat, we can track the biomass of the dead organic matter (D) and the biomass of the raccoon population (B_x).

To keep it very simple, we will assume that only humans produce the dead organic matter and that this matter is the only food source for the raccoons. Let us assume that the influx of dead organic matter is at a constant rate α (i.e. humans are always producing garbage in a park) while the efflux of dead organic matter is proportional to amount of dead organic matter βD (β is a rate constant that tracks the removal of refuse in the park per unit time) and to the amount consumed by the raccoons cDB_x (c is a rate constant that describes the consumption of raccoons per unit biomass per unit time). The influx of biomass into the raccoons is equal to the amount consumed minus the amount egested, which is determined by the efficiency constant ϵ (dimensionless) multiplied to cDB_x. Finally, the efflux of biomass out of raccoons is equal to the proportion of raccoon biomass lB_x (l is a rate constant that describes the loss of biomass due to respiration (l_r) and mortality (l_m) per unit time (i.e. $l = l_r + l_m$). These assumptions lead us to the following system of ordinary differential equations:

$$\frac{dD}{dt} = \alpha - \beta D - cDB_x \tag{1.2a}$$

$$\frac{dB_x}{dt} = \epsilon cDB_x - lB_x \tag{1.2b}$$

This model is both similar and dissimilar to the logistic growth model presented above. First, we can have a raccoon-free equilibrium, $E_1 = (D_1, B_{x,1}) = (\alpha/\beta, 0)$. The eigenvalues associated with this equilibrium are $\lambda_1 = -\beta$ and $\lambda_2 = \epsilon(\alpha/\beta)c - l$. Since the first eigenvalue is always negative, the second eigenvalue determines the linear stability of the raccoon-free equilibrium and the expression is equivalent to ensuring that the raccoons can maintain a positive biomass at this level of resources. Thus, if losses are too high, resources are too poor in quality and/or quantity or if the attack rate is too low, then raccoons have a negative growth rate (as if r is negative in logistic growth) and go locally extinct. If none of these conditions are the case, then the raccoons can invade the equilibrium. Furthermore, if raccoon biomass is low enough that $dD/dt \approx 0$, then the biomass will increase exponentially at a rate $\epsilon(\alpha/\beta)c - l$, which is similar to the logistic growth model at low densities.

The biomass of the raccoon then increases towards its value found at the dead organic matter raccoon equilibrium, $E_2 = (D_2, B_{x,2}) = (l/(\epsilon c), (\alpha\epsilon c - \beta l)/(cl))$ (Fig. 1.2). However, raccoon biomass commonly exceeds its equilibrium biomass due to overconsumption of its resources (Fig. 1.2). This overconsumption, which is due to the lag in time scales between the resource and consumer, does not lead to sustained oscillations as would occur in a pure consumer-resource population dynamic [168]. This result can be seen through an examination of the Jacobian matrix of E_2 and its eigenvalues:

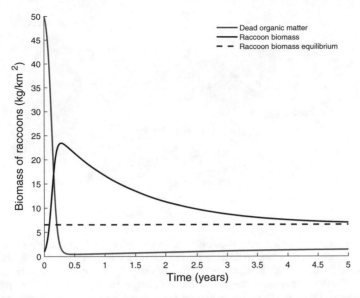

Fig. 1.2 Dynamics of raccoon biomass (solid black line) and dead organic matter (solid blue line) over time. The biomass of the raccoons drastically overshoots the equilibrium biomass, despite the equilibrium being a stable node (i.e. its eigenvalues lack imaginary parts)

$$
\mathbf{J}_{E_2} = \begin{pmatrix} \frac{\partial F_1}{\partial D} & \frac{\partial F_1}{\partial B_x} \\ \frac{\partial F_2}{\partial D} & \frac{\partial F_2}{\partial B_x} \end{pmatrix} = \begin{pmatrix} j_{11} & j_{12} \\ j_{21} & j_{22} \end{pmatrix} = \begin{pmatrix} -\frac{\alpha \epsilon c}{l} & -\frac{l}{\epsilon} \\ \frac{\epsilon^2 \alpha c}{l} - \epsilon \beta & 0 \end{pmatrix} \tag{1.3a}
$$

$$
\lambda_{1,2} = \frac{j_{11} \sqrt{j_{11}^2 + 4 j_{12} j_{21}}}{2} \tag{1.3b}
$$

The first thing to note is that j_{11} and $j_{12} j_{21}$ are negative numbers, which leads to (a) the eigenvalues always having a negative real part and hence E_2 is locally stable, and (b) it is possible to have imaginary components to the eigenvalues if $4 j_{12} j_{21} > j_{11}^2$. Thus, if the conditions allow for an equilibrium to be a stable spiral, we expect dampened oscillations towards the equilibrium over time, but sustained oscillations would not be possible. Therefore, the energy-limited ecosystem is more stable than a pure consumer-resource system due to the constraints on growth for the consumer.

We should also note that the outputs of the two models are also not equivalent, for increases in biomass do not necessarily lead to increases in population size and vice-versa. It is only true if individuals are of a fixed size (i.e. all raccoons weigh 6kg, thus increases in biomass lead to increases in population sizes). This approximation is roughly true with unicellular organisms that reproduce through binary fission or mitosis, or for multicellular organisms capable of reproducing asexually, but in many species, biomass accumulation is somewhat decoupled from sexual reproduction.

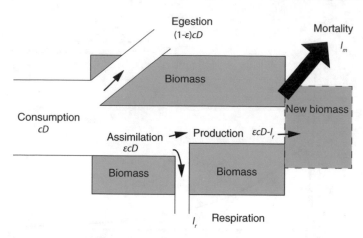

Fig. 1.3 Illustration of how energy is assimilated and turned towards production in the raccoon population as described by the functions in Eq. 1.2. The green box represents the biomass of the population, while the 'channels' represent the flux of energy as it is consumed, then assimilated with part of the flux exiting through egestion. The assimilated energy represents gross production. Net production is the energy that remains after respiration, and is used for growth, reproduction and storage, creating new biomass. This gain in biomass is offset by losses due to mortality. Adapted from [137] with modifications suggested by [9]

This is why when discussing energy in ecosystems, we need to use the concepts of *assimilation*, *production* and *respiration* (Fig. 1.3). Organisms can only assimilate a portion of the energy that they consume, with a significant portion being lost to egestion (Fig. 1.3). The ratio between assimilation and consumption, the assimilation efficiency, is denoted in Eq. 1.2 by ϵ. A portion of this assimilated energy is then utilized for active metabolism (e.g. think of starch and fats being broken down into CO_2) and is respired by the organism, and therefore cannot be used to synthesize new organic molecules. Thus, only non-respired, assimilated energy can be used to new biomass, which is what we call production (Fig. 1.3). Production, under this formulation, is the rate of new organic molecules that are synthesized by an organism and thus contribute to biomass through either growth of individuals, reproduction and/or storage. Two other ecosystem ratios that arise from this are the gross production efficiency (production divided by consumption) and the net production efficiency (production divided by assimilation) as they indicate how well energy is capture by a population, community or trophic level.

There are other losses to biomass and/or production, though where those losses occur is not always defined the same way. Organisms also excrete and/or secrete organic molecules, with animal energy budget literature considering it a loss from assimilated energy like respiration [9, 107, e.g.] while the plant energy budget literature considering it as a use of production [22]. This difference can be explained in part by the origins of most of the excretions: animal primarily excrete metabolic wastes, while plants excrete exudates that are functional [22]. Similarly, losses due

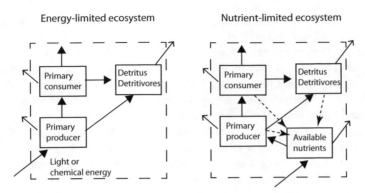

Fig. 1.4 How flows between compartments and the environment differ in energy and nutrient-limited ecosystems. In energy-limited ecosystems, new energy is constantly inputed to counterbalance the losses as every organic compartment loses energy to respiration. In nutrient-limited ecosystems, there is significant nutrient cycling that reduces the importance of new nutrient inputs

to mortality can be modelled as a loss from production [22] or it can be viewed as a loss from biomass [9, 107]. Here, we will generally lump losses after assimilation as a generic mass-loss rate (i.e. l in Eq. 1.2), but it pays to be aware of this assumption as it influences the interpretation of the results.

In our ecosystem model, the production of raccoon biomass is $\epsilon c D B_x - l_r B_x$, which, at equilibrium, is equal to $l_m B_{x,2} = l_m (\alpha \epsilon c - \beta l)/(cl))$ as inputs are equal to outputs (Fig. 1.3). Thus, when production equals losses due to mortality, our raccoon biomass no longer changes as fluxes into the biomass equal the fluxes out. This type of production is known as secondary production, as raccoons are heterotrophs and reliant on food. At the base of energetic ecosystems, we usually find primary producers that turn light or chemical energy into complex molecules (Fig. 1.4). Assimilated energy for primary producers is called *gross primary production*, while production after respiration is know as *net primary production*. In energy-limited ecosystems, the productivity of primary producers along with the efficiencies of the various biotic ecosystem compartments determines the amount of total biomass and ecosystem production [22].

However, energy is not the only or even the primary limiting factor in most ecosystems. Instead, nutrients are commonly controlling the growth of biomass. For example, there is strong evidence that phosphorus limits the growth of algae blooms in temperate lakes [158]. One of the major differences between the fluxes of nutrients and the fluxes of energy in an ecosystem is that energy typically follows a linear path while nutrients cycle [36, 107]. Furthermore, most energy that enters an ecosystem is lost, while limiting nutrients can be tightly conserved. Nutrient-limited ecosystems can be functionally closed when there are limited inputs and losses are minor, while energetically-limited ecosystems need to be opened in order to be sustained (Fig. 1.4).

The practical upshot of these differences can be reflected by looking at a 'minimal' ecosystem model that many others have used before [35, 64, 107]. This ecosystem has two ecosystem compartments: an available pool of a limiting nutrient N which is taken up directly by a primary producer who has biomass B_x. Nutrients are supplied to the ecosystem at a constant rate I_N and are lost to the pool through a proportional efflux rate ($E_N N$) as well as uptake by the primary producer $c B_x N$ (c is the uptake rate of the limiting nutrient per unit biomass).

To convert limiting nutrients into biomass, we introduce the concept of a quota, where Q_N is proportion that N takes up in terms of biomass of the primary producer. For example, if B_x is 100 grams and N within the biomass is 4 grams, then $Q_N = 4/100 = 0.025$. In many ecosystem models, this quota will be fixed (i.e. $Q_N = q_N =$ constant) and this leads one to model the nutrient stock inside the primary producers (and higher trophic levels) directly, i.e. $N_x = B_x q_N$ [35, 107]. For this example, we will stick with modelling the biomass directly, but we will use nutrient stocks elsewhere in later chapters. Therefore, biomass of the primary producer will increase by $c B_x N / q_N$, as we assume perfect assimilation of nutrients. We also assume that primary producer biomass is lost a rate proportional to its biomass, $l B_x$, where l is a rate constant. The nutrients in the lost biomass are then partially or completely recycled back into the available nutrient pool instantaneously, with χ_N being the proportion recycled. These assumptions give rise to the following system of ordinary differential equations:

$$\frac{dN}{dt} = I_N - E_N N - c B_x N + \chi_N l q_N B_x \tag{1.4a}$$

$$\frac{dB_x}{dt} = \frac{c B_x N}{q_N} - l B_x \tag{1.4b}$$

To begin the analysis of this set of equations, let us create a variable, T_N, that keeps that of the amount of nutrient within the whole ecosystem, i.e. $T_N = N + q_N B_x$. The rate of change of total nutrients is then equal to:

$$\frac{dT_N}{dt} = \frac{d}{dt}(N + q_N B_x) = \frac{dN}{dt} + q_N \frac{dB_x}{dt} = I_N - E_N N - (1 - \chi_N) l q_N B_x \tag{1.5}$$

With this equation, let us consider the special case where l is equal to $E_N / (1 - \chi_N)$. When this holds, we have:

$$\frac{dT_N}{dt} = I_N - E_N N - \frac{(1 - \chi_N) E_N}{1 - \chi_N} q_N B_x = I_N - E_N T_N \tag{1.6}$$

The solution to this linear first-order ordinary differential equation is relatively simple:

$$T_N(t) = \frac{I_N}{E_N} + \left(T_N(0) - \frac{I_N}{E_N} \right) e^{-E_N t} \tag{1.7}$$

Total nutrients in the ecosystem therefore converges towards I_N/E_N as time goes to infinity, independent of initial conditions. Therefore, in the asymptotic limit, $N + q_N B_x = I_N/E_N$. At the limit, we can substitute $I_N/E_N - q_N B_x$ for N and eliminate the dN/dt equation, which gives us the following expression for primary producer biomass:

$$\frac{dB_x}{dt} = \left(c \frac{\frac{I_N}{E_N} - q_N B_x}{q_N} - l \right) B_x = \left(\frac{cI_N}{q_N E_N} - l \right) B_x - cB_x^2$$

$$= rB_x \left(1 - \frac{B_x}{K} \right) \tag{1.8}$$

where $r = (cI_N)/(q_N E_N) - l$ and $K = r/c$. Thus, with a small amount of manipulation, we can recapture our logistic growth equation, but this time for biomass instead of population density. We were therefore able to prevent overconsumption because the amount of nutrients became constrained in the ecosystem. To be more precise, with our careful selection of l value, we imposed a *mass balance constraint*, which is normally achieved by fixing the amount of nutrients for all time, i.e. the ecosystem was *closed*. This can be seen readily by fixing T_N to be equal to S, a constant, and then substituting $S - q_N B_x$ for N in the dB_x/dt equation, with the only difference being S replacing I_N/E_N. Thus, proportion of recycling can be key in determining the behaviour of open ecosystems, which is not something we expect in energetically-limited ecosystems.

Of course, the losses of nutrients from the system may not be so perfectly balanced with the recycling of nutrients. If we restrict our attention to the equilibrium where primary producers persist, we have the following:

$$E_p = (N_p, B_{x,p}) = \left(\frac{lq_N}{c}, \frac{I_N - \frac{E_N lq_N}{c}}{(1 - \chi_N)lq_N} \right) \tag{1.9}$$

Thus, equilibrium available nutrient stock is independent of nutrient parameters and is determined solely by primary producer nutrient losses divided by the consumption rate. This occurs because the nutrients are under *recipient-control*, i.e. the recipient of the available nutrients, the primary producer, determines the amount of nutrients. The primary producer biomass is determined by the net supply of available nutrients divided by rate of nutrients lost by the primary producers to the outside of the ecosystem. If the primary producers do not lose any nutrients, there is no possible equilibrium as primary producer biomass goes towards infinity.

The net primary production at equilibrium is simply $lB_{x,p} = \frac{I_N - E_N N_p}{(1-\chi_N)q_N}$, if we assume l captures all post-respiration losses of biomass. Thus, net primary production in this simple ecosystem is equal to the ratio of nutrient supply ($I_N - E_N N_p$) divided by the efficiency of primary producer's retention of nutrients ($(1 - \chi_N)q_N$) [107]. Therefore, the presence of nutrient recycling is an important factor in understanding ecosystem functioning and structure.

1.2   From Ecosystems to Meta-Ecosystems: Layout of the Book

At this point, we hope the reader is comfortable with ecosystem ecology terminology and concepts along with their expression through relatively simple models. We have shown how models of population dynamics and ecosystem dynamics can resemble one another while also having important differences. In particular, the possible feedbacks that exist with the presence of recycling make nutrient-limited ecosystems highlight the importance of nutrient flows from all trophic levels, instead of a focus on linear flows of energy.

However, we have left out a key flow from our presentation of ecosystems: spatial flows. We said our ecosystems were 'opened' or 'closed', but where were the nutrients and/or energy coming from? In the current formulations, we have constant fluxes of energy and/or nutrients into a focal ecosystem and proportional fluxes out. Such constant fluxes make sense for atmospheric deposition or solar energy, but much energy and matter comes from surrounding ecosystems in the landscape. A series of lakes may exchange zooplankton, phytoplankon and nutrients due to connections through groundwater and streams. A forest may receive an important influx of nitrogen and phosphorus from emerging aquatic insects in a stream that runs through it [101]. An estuary may receive massive inputs of nutrients from upstream farms, leading to spikes in primary productivity.

Furthermore, organisms can couple ecosystems through their movements and consumption patterns [124]. A bear can add its fecal matter and urine to a stream while removing salmon from it, and then transport the remnants of the salmon carcass into the forest. Such bidirectional flows of organisms, energy and matter can vary in time and space, and they cannot be easily integrated in standard ecosystem models. Instead, we have to expand our ecosystems into a system of ecosystems or a 'meta-ecosystem' [56, 110, 122].

Over the next four chapters, we will present the conceptual apparatus for meta-ecosystems along with a variety of models to show the power and limitations of this approach. In Chap. 2, we present how meta-ecosystem theory emerged from metapopulation and metacommunity theory along with simple equilibrium analyses that show the impacts of spatial flows across ecosystem boundaries. In Chap. 3, we go beyond equilibrium analyses to look at how meta-ecosystems can vary across time and space, and how these spatiotemporal dynamics can alter ecosystem functioning and structure. In Chap. 4, we show how meta-ecosystem theory can be expanded to include different spatial configurations of ecosystems, species diversity, multiple limiting nutrients and non-resource materials. In Chap. 5, we look at what remains to avenues of future work are being explored in meta-ecosystem theory, including its future applications to ecosystem management.

Chapter 2
Meta-Ecosystems

Abstract Early ecosystem theory, which emphasized strong boundaries between ecosystems, did not consider spatial fluxes in their major explanatory framework based on ecological energetic. To address this limitation, ecosystem ecologists introduced the concept of the landscape, which viewed ecosystems as a continuous mosaic across space. Unfortunately, the continuous mosaic view is difficult to conceptualize and analyze mathematically and is usually restricted to a specific spatial scale of analysis. One potential solution to the problem was to simplify concept by viewing ecosystems being constructed out of discrete patches, which was a major conceptual development in population biology and can be recommended in a number of ecological systems. Therefore, space matters primarily in determining the existence of ecosystem boundaries and the possibility of fluxes across those boundaries. Furthermore, the dynamics of ecosystems, if they are dominated by the flux of materials rather than energy, can be simplified with conservation laws. These simplifications allow us to then talk about a meta-ecosystem, which is this set of discrete ecosystems connected through the fluxes of organisms, materials and energies across ecosystem boundaries. We build on a brief review of metapopulation and metacommunity dynamical models to present a minimal meta-ecosystem model. We demonstrate some important insights derived from early meta-ecosystem models proposed by Loreau et al. [109, 110] to motivate the use of the framework. We then present later work that has shown how flows of matter can further our understanding of species interactions, species assembly, and of ecosystem response to environmental fluctuations.

2.1 A Discrete-Space Approach to Spatial Dynamics

Ecosystem theories have been built on the notion that ecosystem dynamics can be understood as the coupling between discrete ecosystem compartments through fluxes of matter and energy. Ecosystem theory integrated fluxes of energy and matter across large-scale ecosystem boundaries. Later ecosystem dynamics models emphasized more detailed and mechanistic description of local cycling of matters across biotic and abiotic compartments. The arbitrary classification of ecosystems into discrete

© The Author(s), under exclusive license to Springer Nature Switzerland AG 2021
F. Guichard and J. Marleau, *Meta-Ecosystem Dynamics*, Lecture Notes
on Mathematical Modelling in the Life Sciences,
https://doi.org/10.1007/978-3-030-83454-8_2

compartments can correspond to observed discontinuities between functional groups (e.g. primary producers) or ecosystem types (e.g. forest vs lakes). Although subjectively defined, discrete ecosystem compartments facilitated the assessment of coupling in natural systems, by assuming such coupling can be defined at macroscopic scales corresponding to large amounts of matter, rather than at the level of individuals. Spatial dynamics has taken a similar approach with metapopulation theory of population movement between spatially discrete habitats. The discretization of ecosystem components (individuals and matter) and of landscapes is far from being the only approach to the spatial dynamics of ecosystems. Ecologists have long used continuous space formalism to model spatial dynamics [18, 138, see for reviews]. Reaction-diffusion models can potentially allow for more realistic implementation of fine-scale processes causing spatial fluxes, and they also make a number of tools available for the mathematical analysis of models. Representing landscapes as sets of discrete habitats connected by spatial fluxes of individuals and matter provides a powerful simplification for the complexity of natural landscapes over broad ranges of scales. It has also been central to the integration of so many insights gained from metapopulation and metacommunity models into meta-ecosystem or metaecology [157, sensu] theories.

2.2   From Metapopulations to Meta-Ecosystems

Population biology has a long history of studying the role of individual movement for the spatial propagation of populations and for the maintenance of spatial heterogeneity in the distribution of abundance. Spatially discrete spatial models have been introduced to predict the role of limited movement for the persistence of population over landscapes. Spatially distinct local populations were introduced in these metapopulation models to track the proportion of non-extinct local populations resulting from the balance between local population extinction and their recolonization from other local populations [103]. These dynamic theories have been further developed and became popular for their applicability to natural populations. They later integrated spatial movement of multiple species and of matter by extending metapopulation and metacommunity theories to a meta-ecosystem framework.

2.2.1   Scaling-Up Local Ecological Processes to Landscapes

The dynamics of populations over whole landscapes starts from the assumption that ecological interactions among individuals occur over small (local) spatial scales. Over much larger scales (landscapes), it is instead the limited movement of individuals through migration and dispersal of propagules that can affect dynamics. Metapopulation theory assumes that landscapes can be described as a set of spatially discrete local populations connected by the movement of individuals. In its simplest form, metapopulation models only track the occupancy of local populations through population extinction and colonization from other occupied populations. With all-to-all (spatially implicit) connections among local populations, changes in the proportion p of occupied populations is:

$$\frac{dp}{dt} = cp(1 - p) - mp, \tag{2.1}$$

where c is the colonisation rate from occupied populations and m is the extinction rate. At equilibrium. $p^2 = 1 - m/c$ and metapopulation persistence requires that $c > e$. This basic metapopulation model has been extended to multiple species by Hastings [73], Nee and May [133], Tilman [165]. Multi-species metapopulations models have then been extended to metacommunity theories by explicitly tracking local species abundance rather than species occupancy [131, 132]:

$$\frac{dP_{ik}}{dt} = V_k \sum_{l=1}^{N} b_{ilk} P_{il} - m_{ik} P_{ik}, \tag{2.2}$$

with

$$V_k = 1 - \sum_{j=1}^{S} P_{jk} \tag{2.3}$$

where P_{ik} is now the abundance of species i in habitat k, b_{ilk} is the colonization rate of individuals of species i from all other habitats l to habitat k, and m_{ik} the rate of loss of individuals of species i from habitat k. Net change in species abundance is proportional to habitat V_k available for species i given the abundance of all other S species. Metacommunity models were originally developed as a contribution to competition theories at a single trophic level to predict the assembly of species from a combination of local competitive, spatial, and stochastic processes. Metacommunity models have also been central in the study of neutral versus niche theories of species diversity and relative abundance [23].

One of the main predictions of metacommunity theory is the regional distribution of species diversity is based on both local competition and regional movement of species. In competitive metacommunity models, sorting (or environmental filtering) is the local process that operates through competitive exclusion, and can lead to regional patterns of diversity through spatial heterogeneity in the environment. Movement among local communities is the regional process that redistributes, and eventually homogenizes local communities, working against local sorting. Sorting is purely local in the absence of movement, while strong movement leads to regional sorting that favours the best competitor in the average habitat [131]. Here we are adopting spatially-implicit models where each habitat is not given a unique (explicit) spatial position or relationship to other habitats, thus assuming a homogeneous (all-to-all) connectivity network. Under this assumption, intermediate movement reveals the reciprocal effects of local and regional processes: both regional and local diversity and relative species abundance then result from the balance between local sorting and immigration from the regional metacommunity. This balance can for example predict the contributions of both local species richness (alpha diversity) and regional species turnover (beta diversity) to community stability, through local compensatory effects, and to the maintenance of regional asynchrony in species composition [174].

2.2.2 Meta-Ecosystem Models

Spatial ecosystem dynamics has been studied, long before meta-ecosystem models were introduced, as part of early ecosystem ecology through flows of matter across macro-habitats and through large-scale quantitative ecosystem models [66, 145]. However, the meta-ecosystem concept has emerged as a more specific integration of competitive metacommunities and landscape ecosystem ecology [121]. It builds on the study of spatial subsidies from donor to recipient ecosystems [2, 81, 100, 145, 163] to consider feedbacks between donor and recipient ecosystems. As such, meta-ecosystem theory integrates the cycling of matter across scales, and can directly predict local versus regional control of community structure and ecosystem functions through the interplay between local recycling and regional fluxes of matter. The first integration of ecosystem dynamics to metacommunity theories was proposed by [110] who defined the size X_{ik} of an ecosystem compartment i in habitat k and its change through local cycling G_{ij} and spatial flow F_{ilk} to any other ecosystem compartment j and habitat l. In a closed meta-ecosystem we impose a mass balance to matter stored in biotic and abiotic ecosystem compartments as follows

$$\frac{dX_{ij}}{dt} = \sum_k F_{ikj} - \sum_k F_{ijk} + G_{ij}, \qquad (2.4)$$

with local mass balance resulting in:

$$\sum_i G_{ij} = 0 \qquad (2.5)$$

This closed meta-ecosystem can be opened to external inputs and outputs to and from biotic and abiotic compartments. However, imposing a mass balance constraint provided an interesting perspective on spatial dynamics of matter where spatial fluxes are required for local growth to be achieved at the expense of other habitats. More specifically [110] defined

$$\sum_{i,k} (F^*_{ilk} - F^*_{ikl}) = 0 \qquad (2.6)$$

as the net spatial source-sink flow between habitat k and l at equilibrium (denoted by $F*$). These equilibrium constraints can for example result in net nutrient flows in one habitat going from nutrients to consumers (net production), and from consumers to nutrients (net recycling) in the other habitat. This asymmetry is maintained by net spatial fluxes of nutrients where the net recycling habitat becomes the nutrient source for the net producer habitats.

In addition to mass balance constraint, the meta-ecosystem framework also redefined ecological interactions among ecosystem compartments at a more general level and primarily based on trophic interactions where nutrients stocks can be transferred between compartments from resources to consumers. Dynamics are also based on recycling where nutrient stocks stored in biotic compartment can become available to primary producers. The meta-ecosystem framework is a generalization of earlier metacommunity models, and early meta-ecosystem theory has been developed based on linear models and equilibrium dynamics defined above, but with additional ecosystem compartments corresponding to known ecological systems and interactions. These models have revisited metapopulation and metacommunity theories of species coexistence under the conservation of mass through the movement and recycling of nutrient stocks. Meta-ecosystem models offer the possibility of describing spatial fluxes of matter between different ecosystems that might even have no overlap in species distribution, which is not possible when describing spatial fluxes of species in metapopulation and metacommunity models. However, most current meta-ecosystem theory is still built on fluxes between ecosystems of the same type (sensu [122]).

2.3 Insights from Simple Meta-Ecosystem Models

Equilibrium approaches to the meta-ecosystem perspective provides a powerful theoretical framework to address population-level to ecosystem-level questions in spatial ecology including those related to heterogeneous landscapes, to the large-scale assembly of competing species and to external subsidies of organic and inorganic matter. Simple meta-ecosystem models such as Eq. 2.4 assume linear flows between ecosystem compartments and between local ecosystems, resulting in stable steady (equilibrium) states. This assumption of linearity has important impacts on the range of predictions that are possible from meta-ecosystem dynamics. In metapopulation models, linearity comes from constant per capita colonization and extinction rates Eq. 2.1, and resulting equilibrium dynamics predicts a steady state occupancy of populations within the metapopulation. This limits metapopulation models to predict persistence, as far as qualitative predictions are concerned, since the original metapopulation model far too simplistic to make realistic quantitative predictions. Qualitative predictions can be expanded to community-level steady state properties such as diversity and relative abundance distributions in metacommunity models.

At the ecosystem level, and as we saw above and in Chap. 1, new steady state properties can be predicted involving non-living compartments. Rather than a complete shift from metacommunity models, meta-ecosystem models can now use non-living compartments as mechanisms underlying population and community-level predictions. For example, and building on their net spatial fluxes as a mechanism for local growth under regional mass balance (Eq. 2.6), linear meta-ecosystem models have been used to predict the role of spatial flows of matter for the emergence of positive interspecific interactions between species competing for a single resource [57]. These predictions require the extension of the minimal meta-ecosystem model (Eq. 2.14) to consider the dynamics of both organic and inorganic ecosystem compartments, and of multiple trophic levels (in addition to primary producers).

2.3.1 Emergence of Facilitation Among Competitors

The net spatial source-sink flow from Eq. 2.6 can be extended to flows of a single inorganic nutrient (N), a primary producer (P), an herbivore (H), and the detritus (D) resulting from mortality in living ecosystem compartments. Meta-ecosystem dynamics can then be represented by the following differential equations:

$$\frac{dN_x}{dt} = I_x - e_N N_x - f_{Px}(N_x, P_X) + r(1 - e_D)D_x + \Delta_{Nx} \tag{2.7a}$$

$$\frac{dP_x}{dt} = f_{Px}(N_x, P_X) - m_P P_x - f(Hx(P_x, H_x) + \Delta_{Px} \tag{2.7b}$$

$$\frac{dH_x}{dt} = f_{Hx}(P_x, H_x) - m_H H_x + \Delta_{Hx} \tag{2.7c}$$

$$\frac{dD_x}{dt} = m_P P_x + m_H H_x - r D_x + \Delta_{Dx}. \tag{2.7d}$$

where the subscript x refers to a patch. The local inorganic nutrient compartment is open to external input at rate I_x and to output at rate e_N. At equilibrium, the inflows balance the outflows as imposed by global mass balance. The consumption of the nutrient by the producers is given by the function f_{Px} and the consumption of the producers by herbivores is f_{Hx} that are assumed to be linear. Producers and herbivores die at the density-independent rates m_P and m_H, respectively, and their biomass is completely incorporated into the detritus compartments. The nutrient is recycled from the detritus compartment at rate r (mineralization rate) and a fraction e_D is lost during this process. In the absence of spatial flows and consumption, the inorganic nutrient concentration is I_x/e_N, a quantity we can refer to as patch fertility. Spatial flows between patches for each ecosystem compartment is modeled using general dispersal functions Δ_{Nx}, Δ_{Px}, Δ_{Hx}, and Δ_{Dx}, for the nutrient, producers, herbivore, and detritus, respectively.

Inorganic nutrient uptake by producers is described by the functional response $f_{Px} = \alpha_x N_x P_x$, where α_x is the producer consumption rate. The producer reaches steady state at a minimum inorganic nutrient concentration $N_x^* = m_P/a_x$. For the herbivore, we can consider both donor control ($f_{Hx} = \beta_x P_x$, where β is the herbivore consumption rate) and recipient control ($f_{Hx} = \beta_x P_x H_x$) linear functional responses.

We further assume that dispersal is passive, with no preferential movement to any patch. For a given compartment C in a meta-ecosystem of two patches (patch 1 and patch 2), spatial flow is defined as $\Delta_{C1} = d_C(C_2 - C_1)$ for patch 1 and the opposite for patch 2.

Metacommunity theories can predict heterogeneous species dominance among patches from limited species dispersal. This prediction is important because it shows the important of spatial fluxes of individuals, in addition to spatial environmental heterogeneity among habitats, to explain spatial heterogeneity in local relative abundance of species. Meta-ecosystems can provide a more mechanistic representation of spatial environmental heterogeneity through the recycling and spatial fluxes of nutrients stocks. As metacommunity models, some parameters such as fertility can be specific to local patches x. However, other properties such as nutrient stocks N_x^*, that would affect metacommunity dynamics implicitly as fixed parameters can now be explicitly defined as state variables and vary among locations. Let's consider a specific example where the two patches have the same fertility, but the environment varies so that the primary producer could not maintain a population in patch 2. That could happen if environmental conditions in patch 2 (other than nutrient supply determining fertility) impose a higher mortality or lower growth rate than in patch 1. At equilibrium and with no spatial flows of any kind, the nutrient concentration will be lower in patch 1 than in patch 2 because the primary producer depletes nutrients in the only patch (patch 1) where it can persist. Now if we allow spatial flows of the inorganic nutrient and biomass between the patches, the inorganic nutrient will flow from patch 2 to patch 1, while the nutrient sequestered in the biomass will flow in the opposite direction.

The fundamental density-independent per capita population growth rate of the producer is again patch specific and is independent from spatial fluxes. It is given by:

$$\lambda_x = \frac{I_x}{e} - N_x^* \tag{2.8}$$

The realized density-independent *per capita* population growth rate of the producer in patch 2 when coupled via nutrients to patch 1 is defined by the local environment and by the neighboring ecological context. It is given by:

$$\lambda_{21} = \hat{N}_2 - N_2^* \tag{2.9}$$

where $\hat{N}_2$ is the equilibrium inorganic nutrient concentration in patch 2 when coupled to patch 1 in absence of the primary producer in patch 2. The difference between Eqs. 2.8 and 2.9 gives the net effect of the spatial flows (meta-ecosystem effect) of the nutrient on the density-independent population growth rate:

$$\Phi_{21} = \hat{N}_2 - \frac{I_2}{e_N} \tag{2.10}$$

In the simplest case corresponding to Eq. 2.4 where we have only an inorganic nutrient and a primary producer, the equilibrium inorganic nutrient concentration in patch 2 is then

$$\hat{N}_2 = \frac{I_2 + d_N N_1^*}{e + d_N} \tag{2.11}$$

and the meta-ecosystem effect is

$$\Phi_{21} = d_N \frac{N_1^* - \frac{I_2}{e}}{e + d_N} \tag{2.12}$$

There are two possible consequences of spatial flows of the inorganic nutrient between patches. First, the inorganic nutrient could flow from patch 2 to patch 1 if the producer in patch 1 depletes the nutrient to lower concentration than the fertility of patch 2 ($N_1^* < I_2/e_N$). This negative effect of spatial nutrient flows on inorganic nutrient availability in patch 2 could be sufficient to transform it into a sink if it makes the realized growth rate negative ($\lambda_{21} < 0$). Alternatively, if the producer in patch 1 depletes the inorganic nutrient to higher concentration than the fertility of patch 2 ($N_1^* > I_2/e_N$), the inorganic nutrient flows from patch 1 to patch 2 and enriches the latter. The magnitude of this effect scales asymptotically with d_N and could turn a sink into a source (when $\lambda_{21} > 0$). A net flow of the nutrient from patch 1 (the source) to patch 2 is, however, detrimental to population size in patch 1 and can prevent the persistence of the primary producer even if patch 1 is a source according to its fundamental growth rate.

In a system with a detritus compartment in which both nutrient and detritus can flow between patches, the effect of the spatial flows on the inorganic nutrient concentration in the patch 2 could be positive or negative, depending on the difference of fertility between patches. In this more complex case, the meta-ecosystem effect Φ_{21} can still be assessed by solving $\Phi_{21} = 0$ for d_D to find the critical detritus spatial flow d_D^{crit} distinguishing a positive from a negative effect:

$$d_D^{crit} = \frac{r e_D d_N \left(\frac{I_2}{e_N} - N_1^*\right)}{e_N(1 - e_D)\left(\frac{I_1}{e_N} - N_1^*\right) - 2e_D d_N \left(\frac{I_2}{e_N} - N_1^*\right)} \tag{2.13}$$

If $d_D > d_D^{crit}$, the primary producer in patch 2 will benefit from the presence of the primary producer in patch 1. If patch 2 is a sink, it becomes a source if spatial flows increase nutrient concentration to values above the fundamental growth rate ($\lambda_{21} > 0$). Similarly, if patch 2 is a source, it becomes a sink if spatial flows decrease the inorganic nutrient concentration below the fundamental growth rate ($\lambda_{21} < 0$). An essential aspect of Eq. 2.13 is that the critical spatial flux of detritus depends on the relationship between within-patch fertility values (I_1 and I_2; Fig. 2.1) and on the tolerance of primary producers to low resource level in patch 1 (N_1^*; Fig. 2.2). For example, if the fertility is higher in patch 2 ($I_1 < I_2$), then the nutrient is more likely

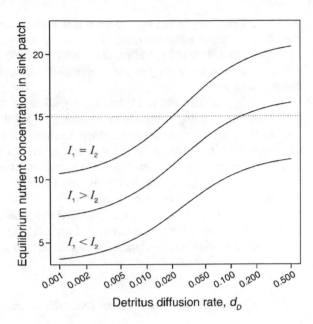

Fig. 2.1 Effect of the detritus flows on the inorganic nutrient concentration in the sink patch for different fertilities (I = inorganic nutrient input rate) in the source (patch 1) and the sink (patch 2) patches. The dotted line shows the fertility (equlibrium nutrient stock in the absence of other ecosystem compartments) in the sink patch (from [57])

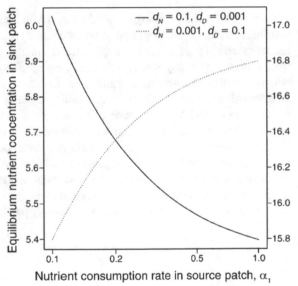

Fig. 2.2 Effect of the primary producer traits (here the consumption rate) in the source patch on the inorganic nutrient concentration in the sink patch (from [57])

to flow from patch 2 to patch 1, thus reducing the inorganic nutrient concentration in patch 2 below its fertility (dotted line on Fig. 2.1). In all three situations illustrated on Fig. 2.1, a net flow of the nutrient from patch 1 to patch 2 is detrimental to the population in patch 1. Population size in patch 1 is reduced by the spatial flow of the inorganic nutrient and this reduction could be sufficient to prevent the persistence of the primary producer in that patch even if it is a source according to the fundamental growth rate. This result illustrates the importance of considering the meta-ecosystem as a whole to understand the impacts of nutrient flows: the net nutrient flow into a patch depends on the fertility and the community composition of the patches within its neighborhood.

The main goal here was to explore the significance of non-living compartments, their coupling to living compartments and their spatial fluxes, for population and community properties such as persistence and coexistence. We learned that habitat heterogeneity is not necessarily an imposed properties of landscapes that can affect species persistence and coexistence. The cycling and spatial fluxes of nutrients and detritus makes species the drivers of this heterogeneity and the engineers of species diversity. As we will see now, this simple meta-ecosystem model can be extended to address questions that are more closely related to its metapopulations and metacommunity predecessors. These questions must include the problem of species assembly from colonization-extinction dynamics over large landscapes.

2.3.2   Colonization-Extinction over Meta-Ecosystems

Meta-ecosystem theories have been developed from metapopulation theories by substituting patch occupancy by populations with the dynamics of nutrient stock as a continuous variable proportional to their nutrient content. Predictions have thus shifted from species occupancy to equilibrium nutrient stock across landscapes. If the goal is still to describe the assembly and structure (*i.e.* relative abundance) of communities, an alternative approach consists in combining dynamics of patch occupancy and of nutrient stocks [58]. We can first describe the dynamics of colonization and extinction of multiple species at the regional scale following the classic metapopulation model Eq. 2.1. However, we can substitute simple parameters of colonization and extinction rates with rates based on the cycling and spatial flows of nutrients and detritus. As we saw above, spatial fluxes of organic and inorganic stocks can determine the ability of species to maintain local population in each local patch. Local ecosystem dynamics can first be extended to both occupied and empty patches from Eq. 2.7a as follows:

$$\frac{dN_{xo}}{dt} = I - e_N N_{xo} - \sum_i a_i N_{xo} B_{xi} + r D_{xo} - d_N (N_{xo} - \bar{N}) \quad (2.14a)$$

$$\frac{dN_{xe}}{dt} = I - e_N N_{xe} + r D_{xe} - d_N (N_{xe} - \bar{N}) \quad (2.14b)$$

$$\frac{dB_{xi}}{dt} = a_i N_{xo} B_{xi} - b_i B_{xi} \quad (2.14c)$$

$$\frac{dD_{xo}}{dt} = \sum_i b_i B_{xi} - r D_{xo} - e_D D_{xo} - d_D (D_{xo} - \bar{D}) \quad (2.14d)$$

$$\frac{dD_{xe}}{dt} = -r D_{xe} - e_D D_{xe} - d_D (D_{xe} - \bar{D}) \quad (2.14e)$$

where $\bar{N} = pN_{xo} + (1 - p)N_{xe}$ and $\bar{D} = pD_{xo} + (1 - p)D_{xe}$ are regional averages of nutrients and detritus respectively. The plant biomass of species i in patch x is B_{xi}. The inorganic nutrient and detritus compartments in occupied (index o) and empty (index e) patches are N_{xo} and N_{xe} and D_{xo} and D_{xe}, respectively. The meta-ecosystem is open to nutrient inputs and outputs. The inorganic nutrient compartment in each local patch receives external inputs (e.g., atmospheric depositions, rock alteration) at rate I and is exported out of the meta-ecosystem at rate e_N (e.g., in-depth nutrient leaching). The plants consume the inorganic nutrient in a patch at rate a_i and die at rate b_i. Dead plant tissue (e.g., foliage) and individuals are completely incorporated into the local detritus compartment. The detritus is mineralized at rate r. A fraction of the detritus is lost from the ecosystem at rate e_D.

The inorganic nutrient availability in empty patches is a critical quantity that determines the ability of an inferior competitor to invade empty patches and thus coexist with resident species. It also informs us on the direction of the net flow between occupied and empty patches. A net flow toward empty patches results in an increased equilibrium nutrient concentration in empty patches ($\hat{N}_{xe}$) above its expected value in absence of spatial flows (i.e. $\hat{N}_{xe} > I/e_N$). For a plant species to persist in a single patch, the nutrient concentration in an empty patch must be $N_i^* = b_i/a_i$, and $N_i^* < \hat{N}_{xe}$ is the condition for an empty patch.

This model assumes spatial fluxes are limited to nutrients and detritus, and as with previous models presented so far, these spatial fluxes are passive (diffusive). The difference between nutrient concentrations in empty patches in the presence and in the absence of spatial fluxes tells us whether the spatial fluxes enrich or impoverish empty patches. This transition can be formulated as a critical ratio of diffusion rates, $(d_D/d_N)^*$. We first set $\hat{N}_{xe} - I/e_N = 0$ corresponding to the difference between the equilibrium nutrient concentration in empty patches in the presence ($\hat{N}_{xe}$) and in the absence (I/e_N) of spatial fluxes. We then solve this for d_D/d_N, yielding

$$\left(\frac{d_D}{d_N}\right)^* = \frac{-e_D(r + e_D)}{d_N e_D - r e_N} \tag{2.15}$$

This means that the direction of the net flow between empty and occupied patches depends only on the diffusion rates and the recycling efficiency. The criterion in Eq. 2.15 determines whether the net nutrient flow is from occupied to empty patches or vice versa. More specifically, empty patches benefit from a net nutrient enrichment When $(d_D/d_N) > (d_D/d_N)^*$. Because the spatial occupancy has an important effect on the properties of both occupied and empty patches, we can expect a strong feedback between nutrient dynamics and patch dynamics.

We can now explore the feedback between net flows of nutrients and patch dynamics by relaxing the assumption of p as a fixed parameter to extend the meta-ecosystem model to the patch-dynamical perspective of metapopulation and metacommunity theories discussed above. We first relate among-patch nutrient flows (d_N and d_D) to metapopulation and metacommunity properties by applying metapopulation dynamics from Eq. 2.1 with a colonization rate $c' = cB$ where c is the number of seeds produced per unit biomass and B is the plant biomass:

$$\frac{dp}{dt} = c'p(1-p) - mp, \tag{2.16}$$

According to local ecosystem dynamics, a species can invade a single patch at some equilibrium biomass B_{inv} if $N^* < I/e_N$. The species can then colonize new patches as a function of its seed production cB, and will persist as a metapopulation if $c > m/B_{inv}$. We can solve nutrient dynamics (Eq. 2.14a) for B_{inv} by assuming that p at invasion is negligible, resulting in:

$$\hat{B}_{inv}^* = \frac{(e_N + d_N)(I_a - e_N b)(r + e_D + d_D)}{a m e_N (e_D + d_D)} \tag{2.17}$$

leading to the following critical colonization rate for metapopulation persistence:

$$c^* = \frac{m a b e_N (e_D + d_D)}{(e_N + d_N)(Ia - e_N b)(r + e_D + d_D)} \tag{2.18}$$

The persistence is enhanced when nutrients flow from empty to occupied patches (i.e., $(d_D/d_N) < (d_D/d_N)^*$). This slight addition to Levins' metapopulation model provides us with our first important result for the dynamics of spatially connected ecosystems: metapopulation persistence is promoted by a net flow of nutrients from the empty to the occupied patches.

Consider a two-species situation with a strong competitor (labeled 1) and a weak competitor (labeled 2). If we assume that nutrient dynamics are much faster than regional dynamics, the weak competitor will be able to colonize only vacant patches (with high nutrient levels), while the strong competitor will be able to colonize both empty and occupied patches. The strong competitor will rapidly displace the weak competitor by lowering the nutrient availability in the patches it occupies. The regional dynamics for the two species is then

$$\frac{dp_1}{dt} = c_1' p_1 (1 - p_1) - mp_1 \tag{2.19a}$$

$$\frac{dp_2}{dt} = c_2' p_2 (1 - p_1 - p_2) - mp_2 - c_1' p_1 p_2 \tag{2.19b}$$

A strong competitor will promote the establishment of a weak competitor in an empty patch if it increases the inorganic nutrient concentration of this patch relative to the concentration expected in an unoccupied landscape. The critical d_D/d_N for a facilitative effect is found by solving $\hat{N}_x^{res} - I/e_N = 0$ for d_D/d_N. The nutrient concentration in the empty patch in the presence of the superior competitor will exceed that expected in an unoccupied landscape, provided that $(d_D/d_N) > (d_D/d_N)^*$, with $(d_D/d_N)^*$ given by Eq. (2.15).

The direction of the net flow of nutrients between occupied and empty patches also affects the critical colonization rate of the weak competitor because it affects the biomass of the patch it invades (B_{inv}) and the availability of empty patches ($1 - p_{res}$). The biomass of the weak competitor in the patch where it invades, in

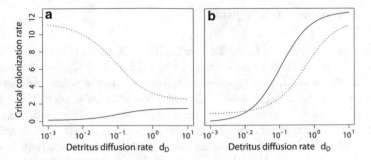

Fig. 2.3 Effects of the ratio of detritus to nutrient nutrient diffusion (shown here as d_D with a fixed $d_N = 0.1$) on regional persistence. **a** Invasion of the inferior competitor in the absence (solid line) and the presence (dotted line) of the superior competitor. **b** Invasion of the superior competitor in the absence (solid line) and the presence (dotted line) of the inferior competitor (from [58])

the presence of a resident strong competitor, is calculated as in Eq. 2.17 and shows how the critical colonization rate for persistence of a weak competitor is modified by d_D/d_N (Fig. 2.3). Colonization by the inferior competitor is always more likely (lower critical colonization rate) in the absence of the superior competitor (Fig. 2.3a, dotted line), but increasing d_D/d_N facilitates colonization. Given that increasing d_D/d_N results in a net nutrient enrichment of empty patches (as seen from Eq. 2.15), we can conclude that coexistence of an inferior competitor is promoted by a net flow of nutrients from occupied to empty patches. Increasing net flows in the direction of empty patches has opposite effects on the superior competitor. However, we can at the same time see that the inferior competitor can facilitate the colonization of the superior competitor by increasing nutrient availability in empty patches (Fig. 2.3b, solid vs dotted line).

The two-species model can be extended to multiple species [58] to show how the emerging facilitation coming from ecosystem dynamics and from the flow of nutrients and detritus can affect the assembly of diverse communities, but also trigger cascades of extinctions when species persistence depends on facilitation associated with meta-ecosystem processes. However, no analytical colonisation criteria are available for more than two species [58]. It is interesting to think about specific processes occurring in natural systems that can control spatial fluxes of nutrients and detritus in a way that can lead to these feedbacks between ecosystem and colonization dynamics. In cropland-forest landscapes for example, large mobile herbivores such as deer can forage in disturbed crop areas that are rich in nitrogen and move these nutrients into mature forests. The meta-ecosystem framework described above could predict how mobile herbivores could drive spatial fluxes of nutrients and slow down forest succession by moving nutrients from early to late succession areas.

2.3.3 Forced Meta-Ecosystems

Examples we covered so far focused on spatial fluxes of matter between local ecosystems that (1) are defined as constant rates in models, and (2) that connect local ecosystems that are of similar type (between forest patches or ponds). However, natural systems that have inspired meta-ecosystem theories are often characterised by ecosystem subsidies in the form of living and non-living organic and inorganic matter that are neither constant nor limited to ecosystem of the same type. This is actually a great strength of meta-ecosystems as an extension of metacommunity models, spatial fluxes of matter, and their impact as ecosystem subsidies are not dependent on fitness in the recipient system. Oak leaves sinking to the bottom of a lake during the fall season can subsidize aquatic primary production irrespective of the fact that oaks don't do well in lakes. We can now think about how connected lake and forest communities are connected through the movement of matter that is blind to ecosystem type. Ecologists have documented the importance of time-varying spatial fluxes of matter in natural systems between distinct ecosystems that otherwise share no or few common species (e.g. aquatic and terrestrial ecosystems [145]).

One other property of such subsidies in natural ecosystems is their temporal variability caused by environmental forcing. Environmental forcing refers to temporal variability in processes that are external to the meta-ecosystem. It can take the form of pulses corresponding to climatic events (rainfalls, storms) or of variability in the movement of organisms acting as vectors for the movement of matter (excretion of nutrients). In this context, forced subsidies can be characterized by their frequency and amplitude [101]. Without yet relaxing the assumption of linear functional responses, meta-ecosystem models can thus become much more applicable to natural systems by studying the role of time-varying ecosystem subsidies between different ecosystem types. One motivation here is to understand how fluxes of matter can contribute to the growth of higher trophic levels in recipient subsidized ecosystems that would not otherwise be coupled through the movement and dispersal of organisms.

Temporally variable subsidies had previously been studied as external forcing that is independent from local ecosystem dynamics, without explicit reference to the source of subsidised matter. Meta-ecosystem theories allows expressing subsidies from an explicit source ecosystem and with forced variability in the movement of matter. For example, Leroux and Loreau [101] studied two local tri-trophic food chains with a nutrient compartment connected by the movement of herbivore biomass, a portion of which contributes to predator growth, while the remaining proportion subsidises the nutrient compartment through recycling (Fig. 2.4). They model subsidies as a rectangular pulse with variable pulse and inter-pulse periods. The rectangular pulse function is defined by a pulse magnitude (z), duration (u), and frequency ($1/f$).

With reciprocal pulsed subsidies, predators switch to feed on herbivore subsidies, resulting in cascading effects in the local and neighboring ecosystem. In both local ecosystem, subsidies result in the initial loss of local herbivores, which cause a

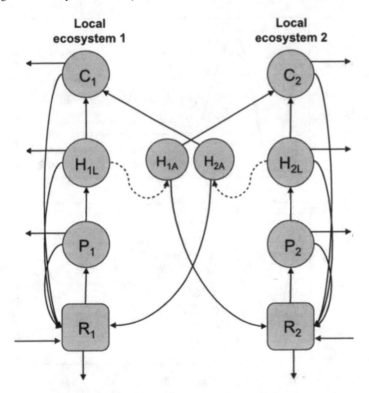

Fig. 2.4 Meta-ecosystem model with allochthonous herbivore subsidies (dashed arrows). In each local ecosystem (1 and 2), contributions to the resource compartment (R) come from external input, recycling of local biotic compartments (P, H_L and C), and through recycling of allochthonous herbivore subsidies coming from the other ecosystem (H_A, dashed arrows). Adapted from [101]

temporary decrease in both herbivore and predator biomass. Predator switching then creates a refuge for the local prey, enabling it to benefit from the presence of the subsidy from the neighboring ecosystem. By switching to feed on herbivore subsidies, predators thus contribute to the increase in neighboring herbivore biomass, which leads to strong cascading effect on producers in the adjacent ecosystem. This is while we still observe a loss of local herbivores via subsidies (from H_L to H_A). The loss of local herbivores to subsidies along with the resulting cascading effect in the recipient ecosystem lead to negative correlation in the biomass of producers between ecosystems despite the fact that producers (P_1 and P_2 in Fig. 2.4) are not directly interacting or competing for the same resource. This phenomenon is known as apparent competition [78] and is here mediated by spatial subsidies at the meta-ecosystem level.

Reciprocal pulsed subsidies further have the potential to generate shifting trophic cascades over short time scales (that is, strength of cascades oscillates between strong and weak, Fig. 2.5). The strength of cascades (PTI_i on Fig. 2.5) in ecosystem i can

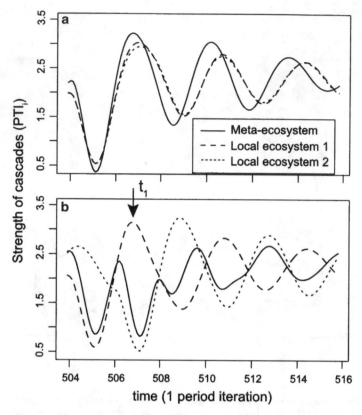

Fig. 2.5 Strengths of cascades in local ecosystem 1 (dashed line), local ecosystem 2 (dotted line), and the meta-ecosystem (solid line), through time when reciprocal pulsed subsidies are (A) in sync and (DŠ) 2 months out of phase. Reciprocal pulsed subsidies lead to asynchronous cascade strength at the local ecosystem scale (for example, see t_1 when cascades are high in local ecosystem 1, they are low in local ecosystem 2) and intermediate strength of cascades in meta-ecosystems (from [101])

be measured with log response ratios: the biomass of producers in local ecosystem i in the presence and the absence of the predator in local ecosystem models. Reciprocal pulsed flows that occur in phase lead to similar temporal signatures in the strengths of local and meta-ecosystem trophic cascades (Fig. 2.5a). Reciprocal pulsed subsidies that occur out of phase lead to cascades of maximum strength that are asynchronous between local ecosystems, resulting in cascades of intermediate strength at the meta-ecosystem scale (Fig. 2.5b). What is important here is that the timing of subsidies, and its relation with the temporal scale of ecosystem dynamics, can affect the magnitude and the spatial heterogeneity (asynchrony) of observed trophic cascades across landscapes. The strength of trophic cascades is typically interpreted as a result of the magnitude of subsidies and of the strength and nature (topology, nonlinearity) of ecological processes such as trophic interactions and recycling.

These predictions are important also because pulsed subsidies have long been documented in many natural systems and form the basis of a number of meta-ecosystem experiments [55, 68, 69] where controlled portions of the organic matter in microcosms can be sampled and exchanged among experimental ecosystems. While subsidies are expected to benefit population growth in recipient ecosystems, these experiments show the limits of such benefits when the frequency or intensity of subsidies driven by disturbances is increased beyond the recovery capacity of source populations [69]. The forcing of meta-ecosystems with periodic pulsed subsidies illustrates the importance of resulting fluctuations in population abundance and species interactions, that can result in strong spatiotemporal heterogeneity across ecosystems (Fig. 2.5). Building on a large body of literature on nonlinear spatial dynamics in ecology, meta-ecosystem theories have been extended to understand the role of spatial fluxes of matter on the endogenous emergence of non-equilibrium spatial dynamics, i.e. in the absence of environmental forcing. This was achieved by integrating density dependence in the form of nonlinear functional responses between trophic compartments.

Chapter 3
Nonlinear Meta-Ecosystem Dynamics

Abstract Early meta-ecosystem theory focused on the balance between movement and recycling of matter assuming linear functional responses and parameter values leading to locally stable equilibrium points. Building on the rich literature on nonlinear dynamics in ecology, more recent studies have shown how nonlinear functional responses interact with movement and recycling to control the spatial and temporal stability of meta-ecosystems and predict the maintenance of strong fluctuations that have important implications for ecosystem persistence and functions. We first review a well-studied nonlinear model of predator-prey dynamics, the Rosenzweig-MacArthur model. We build on this trophic interaction to define a simple two-patch meta-ecosystem and show how the local recycling and between-patch diffusion of nutrients controls the bifurcation to spatially heterogeneous oscillatory dynamics of all ecosystem compartments. We then review how spatial synchrony and principles of weakly-coupled oscillators can be applied to non-equilibrium meta-ecosystems.

3.1 Integrating the Cycling of Matter to Nonlinear Ecological Dynamics

Much of meta-ecosystem theory stems from ecosystem theory that focuses on mass-balanced distribution of matter across ecosystem compartments at equilibrium (steady state). For that purpose, assumptions of linearity in the transfer of matter between compartments has helped develop a set of relationships between meta-ecosystem properties and important dynamical properties of both ecosystem functions and community structure as outlined in Chap. 2. However, a more dynamic approach to ecosystems and more specifically to dynamic feedbacks between trophic interactions and recycling forces us to reconsider the steady state nature of ecosystem functions. Recent studies of non-equilibrium meta-ecosystems point to the importance of recycling and spatial fluxes of matter for the maintenance of endogenous spatiotemporal heterogeneity that can promote both strong local and weak fluctuations. For example, meta-ecosystem models where only nutrients are flowing between local ecosystems lead to the destabilization of local consumer-resource dynamics [115]. Spatial fluxes of matter drive this destabilization, but recycling can facilitate the tran-

© The Author(s), under exclusive license to Springer Nature Switzerland AG 2021   29
F. Guichard and J. Marleau, *Meta-Ecosystem Dynamics*, Lecture Notes
on Mathematical Modelling in the Life Sciences,
https://doi.org/10.1007/978-3-030-83454-8_3

sition from equilibrium to cyclic dynamics. In simple two-patch meta-ecosystems, phase-locked (constant phase difference) cycles emerge from the positive feedback between recycling and passive movement of nutrients: more spatial fluxes into a local ecosystem is correlated with increasing growth of the primary producer, thus destabilizing primary production. This prediction is central to understanding the response to nutrient enrichment of more complex ecosystems that can for example include an organic matter (detritus) compartment [54]. In these cases the top-down control of biomass storage into inorganic form can even stabilize a meta-ecosystem in the face of nutrient enrichment. Non-equilibrium meta-ecosystem theories are also relevant to conservation endeavors by predicting how the emergence of multiple scales of spatiotemporal patterns of abundance can inform decisions on the optimal size and spacing of local protected areas within regional reserve networks [53, 160]. In order to interpret predictions from nonlinear meta-ecosystem models, it is important to first study nonlinear ecological interactions developed in population biology, and understand their impact on the maintenance of spatiotemporal heterogeneity. Meta-ecosystem theory can then integrate these nonlinear trophic interactions and predict how the maintenance of spatiotemporal heterogeneity is impacted by nonlinear processes involving the movement and recycling of matter.

3.2 Nonlinear Dynamics in Space and Time

Theoretical ecology has historically identified the stability of ecological equilibria as one of its main questions, and tied together concepts of stability and of equilibrium (steady) state: constant abundance is the key ecological characteristic of a stable population or community [112]. Stability here is considered in a very broad sense and can be defined in many different ways to fit specific ecological questions [82, 136]. The local stability of fixed points characterizing the equilibrium state of a dynamical system has been one of the main definitions used in theoretical studies. Using this definition, stability has been applied to natural populations as their ability and rate of return to their equilibrium state following a small perturbation. This analysis has pervaded our understanding of population regulation and persistence in heterogeneous environments, of species coexistence, and of community assembly. Almost in parallel, non-equilibrium theories have been developed to show how strong fluctuations, beyond small perturbations of steady-states [39] can lead communities to contrasting patterns of coexistence and exclusion [171] and of population fluctuations [151]. Early studies of stability defined stability as the amount of temporal variation and posited that species persistence is negatively impacted by fluctuations of abundance [112]. In contrast, spatial dynamics theory predicts that the limited movement of organisms can simultaneously increase local fluctuations and promote persistence and coexistence across spatial scales [85, 120].

Non-equilibrium spatial dynamics can result from a combination of intrinsic (nonlinear feedbacks) and extrinsic (environment) forces disrupting steady states of dynamical systems [153]. Although traditional approaches have attempted to partition the relative importance of intrinsic and extrinsic disrupting forces additively [39, see for areview], more recent theories have focused instead on their interaction to explain the maintenance of non-equilibrium spatial dynamics [29, e.g.,]. These theories have been based on a few simple models where non-equilibrium dynamics emerges from implicit time delays in discrete population models [74, e.g. Ricker model;], and/or on nonlinear host-parasitoid and trophic interactions [71, 134]. The Rosenzweig-MacArthur predator-prey model is one of the most common starting point to study non-equilibrium spatial dynamics using ordinary differential systems.

The Rosenzweig-MacArthur predator-prey model describes the nonlinear interaction between predators and their prey and provides a classic example of nonlinear dynamics resulting in positive feedbacks (see below) that lead to non-equilibrium oscillatory dynamics between prey H and predator C biomass:

$$\frac{dH}{dt} = rH(1 - \frac{H}{K}) - \frac{aHC}{b+H} \tag{3.1a}$$

$$\frac{dC}{dt} = \frac{aHC}{b+H} - mC \tag{3.1b}$$

Nonlinear dynamics is here implemented as the saturating (Holling Type II) functional response of predator growth on prey abundance with a maximum capture rate a and half-saturation b. The system has one non-trivial equilibrium point where both H^* and C^* are strictly positive. The stable equilibrium point $\{H^*; C^*\}$ of the system becomes locally unstable and a stable limit cycle emerges through a Hopf bifurcation when

$$K > \frac{b(2b + rm - rb)}{r(b - m)} \tag{3.2}$$

Because the destabilization and the onset of oscillatory dynamics can be driven by increasing carrying capacity of the prey, this dynamical response has been formulated as the 'paradox of enrichment' [154]. The destabilizing effect of the carrying capacity on the predator-prey equilibrium can be generalized to all parameters that increase the flux of biomass from the prey to the predator relative to the loss rate of the predator [152, m in Eq. 3.1]. Interestingly, this energy flux interpretation applies to consumer-resource models with stability defined as the loss of local stability of the equilibrium as well as to stability defined as increasing variability (slower return time) of a perturbed stable equilibrium [152]. In the Rosenzweig-MacArthur Eq. 3.1, the loss of local stability of the steady state with increasing K is associated with increasing temporal variability in population size caused by oscillatory dynamics. More generally, the nonlinear (saturating) *per capita* response of predator growth to prey abundance results in a positive feedback between prey density and prey growth ($\frac{\partial d H/dt}{\partial H} > 0$) at low prey density. This positive feedback destabilizes the equilibrium point and lead to a stable limit cycle. Nonlinear dynamics associated with positive feedbacks is an important driver of non-equilibrium ecological dynamics.

This model can be expanded into a minimal metacommunity model of non-equilibrium spatial dynamics with predator-prey dynamics in 2 discrete habitats, and passive movement (diffusion) of individuals between these habitats:

$$\frac{dH_i}{dt} = rH_i(1 - \frac{H_i}{K_i}) - \frac{aH_iC_i}{b+H_i} + D_H H_j - D_H H_i \qquad (3.3a)$$

$$\frac{dC_i}{dt} = \frac{aH_iC_i}{b+H_i} - m_C C_i + D_C C_j - D_C C_i$$

$$i, j \in \{1, 2\} : i \neq j \qquad (3.3b)$$

This system can display multiple equilibria, some of which are spatially heterogeneous [46, 47]. Non-equilibrium spatial dynamics emerges when any spatially homogeneous limit cycle solution is unstable and is replaced with spatially heterogeneous fluctuations [85, Chap. 4,].

Ecological implications of such non-equilibrium dynamics include the prediction of population persistence and of species coexistence: oscillatory dynamics are predicted to decrease persistence by bringing abundances closer to extinction that can then be caused by perturbations during phases of low abundance. When nonlinear feedbacks lead to self-sustained oscillations, limited movement between habitats can reduce their amplitude or affect their onset (shifting the bifurcation point along control parameters) by decoupling local growth from abundance [14]. In the case of host-parasitoid models, time delays and nonlinear feedbacks predict no persistence in the absence of limited movement. This prediction was tested in the laboratory and provided one of the earliest motivations for studying spatiotemporal dynamics: persistence can be promoted by spatially-asynchronous fluctuations which are themselves maintained by limited movement [72]. In the presence of limited parasitoid movement, implemented explicitly or implicitly, increases in local host abundance can be achieved through passive movement from other locations, which can rescue extinct locations. Such movement can also limit the destabilizing effect of nonlinear feedbacks by locally decoupling growth (e.g., density-dependent growth of the parasitoid) from density. These stabilizing effects can operate as long as locations connected by movement have unequal abundances. In other words, movement must be limited such that local oscillations are not perfectly in phase. These early studies have been extended to other ecological systems characterized by oscillations and prone to extinction. They have led to the general prediction that spatiotemporal dynamics can stabilize and increase persistence of locally fluctuating ecological systems [14] that could otherwise reach extinction. This prediction has been extended to communities [45] and to competitive interactions with the associated problem of species coexistence [25, 139]. However, the role of the recycling and movement of (in)organic matter in this interplay between local and regional fluctuations had remained largely overlooked until more recent meta-ecosystem studies.

3.3 Nonlinear Meta-Ecosystem Dynamics

Integrating the movement and recycling of matter to previous nonlinear models can let us address similar questions of spatiotemporal heterogeneity, but adding the potential effect of fluxes to and from non living compartments on spatial and temporal stability. It also allows extending our range of questions to the role of non-equilibrium dynamics for ecosystem functions in addition to population persistence and species coexistence.

We can now focus on a simple ecological model of two coupled ecosystems [115]. In this model, we concern ourselves with the levels of a limiting nutrient (N) in the medium and the amount of this nutrient bound in the two trophic levels explicitly modeled here, autotrophs (A) and consumers (C) (Fig. 3.1). Nutrient in the biota can be returned to the available pool of nutrient by means of recycling. The general equations describing dynamics in ecosystem i connected to ecosystem j are:

$$\frac{dN_i}{dt} = I_N - E_N N_i + \epsilon_A \delta_A A_i + \epsilon_C \delta_C C_i - f_A(N_i) A_i + d_N(N_j - N_i) \quad (3.4a)$$

$$\frac{dA_i}{dt} = f_A(N_i) A_i - \delta_A A_i - f_C(A_i) C_i + d_A(A_j - A_i) \quad (3.4b)$$

$$\frac{dC_i}{dt} = f_C(A_i) C_i - \delta_C C_i + d_C(C_j - C_i) \quad (3.4c)$$

$$f_A(N_i) = \frac{\alpha_A N_i}{\beta_a + N_i}$$

$$f_C(A_i) = \frac{\alpha_C A_i}{\beta_C + A_i}$$

$$i, j \in [1, 2], i \neq j \quad (3.4d)$$

where ϵ_C, ϵ_A are the proportions of nutrients recycled upon mortality, d_N, d_A, d_C are coefficients of diffusion between ecosystems, δ_A, δ_C are the mortality rates, I_N, E_N are external inputs and loss rates of nutrient, respectively, and f_A, f_C are the functional responses of each trophic level (A and C) to its resource.

We can use this model to study the role of spatial fluxes of nutrients on the onset of non-equilibrium dynamics and spatial heterogeneity. We more specifically control for openness of the meta-ecosystem, which is a measure of connectivity with locations that are beyond the spatial domain of the meta-ecosystem (see Chap. 1). We can also control for connectivity between ecosystems. Each local ecosystem can be open or closed and can be unconnected or connected (Fig. 3.1). The openness of a meta-ecosystem is determined by the values of I_N, E_N, ϵ_C and ϵ_A while the connectivity is determined by the values of d_N, d_A and d_C. If connectivity is not limited, the system is said to be well-mixed and the meta-ecosystem simplifies to a single ecosystem. We therefore investigate four types of meta-ecosystems with respect to openness and connectivity: closed and well-mixed, closed meta-ecosystem, open and well-mixed,

Fig. 3.1 A diagram
describing the various
sub-models of the
nutrient-explicit
meta-ecosystem model.
Symbols are those used in
the equations: consumer and
autotroph abundance (C, A),
nutrient concentration (N),
consumer and autotroph
consumption (f_C, f_A),
autotroph and consumer
mortality rate (δ_A, δ_C),
recycling coefficients $(\epsilon_A,$
$\epsilon_C)$, coefficients of diffusion
(d_N, d_A, d_C) and inputs and
outputs of nutrients $(I_N,$
$E_N)$: **a** Well-mixed, closed,
b Well-mixed, open, **c**
meta-ecosystem, closed and
d meta-ecosystem, open.
Adapted from [115]

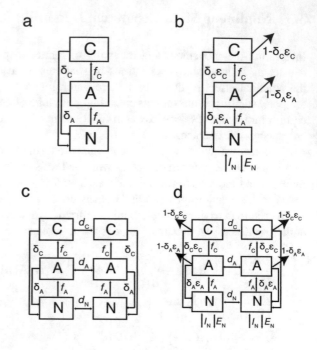

and open meta-ecosystem (Fig. 3.1). For the open and well-mixed meta-ecosystem
(Fig. 3.1b), the equations governing the dynamics are

$$\frac{dN}{dt} = I_N - E_N N + \epsilon_A \delta_A A + \epsilon_C \delta_C C - f_A(N)A \tag{3.5a}$$

$$\frac{dA}{dt} = f_A(N)A - \delta_A A \tag{3.5b}$$

$$\frac{dC}{dt} = f_C(A)C - \delta_C C \tag{3.5c}$$

When the well-mixed ecosystem is closed to external inputs and outputs of nutrients (Fig. 3.1a), the system of differential equations reduces from three to two:

$$\frac{dA}{dt} = f_A(S - A - C)A - f_C(A)C - \delta_A A \tag{3.6a}$$

$$\frac{dC}{dt} = f_C(A)C - \delta_C C \tag{3.6b}$$

For the well-mixed, open ecosystem (Fig. 3.1b), we investigate how differing
sources of ecosystem enrichment (nutrient recycling and external nutrient inputs)
effect ecosystem resilience (i.e. real part of the dominant eigenvalue). We control
total nutrient stock of the ecosystem at equilibrium (S, which is equal to $N + A + C$)
by fixing recycling levels (ϵ_C, ϵ_N) and letting external nutrient inputs (I_N) vary such

Fig. 3.2 Examination of dominant eigenvalues for a well-mixed, open ecosystem with inputs only (solid black), autotroph recycling and inputs (dashed black), consumer recycling and inputs (dashed grey), total recycling and inputs (solid grey). Adapted from [115]

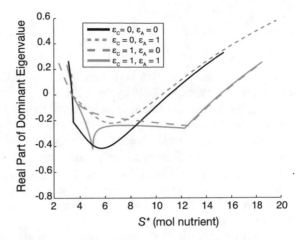

that S is the same across ecosystems. Any differences between the types of recycling are not due to differential allocation of nutrients between compartments. Therefore, any impact of recycling on resilience is due to changes of nutrient flows between compartments. Recycling alters the range of enrichment in which the fixed point can be stable and the resilience of the ecosystem (Fig. 3.2). Consumer recycling decreases stability for low values of enrichment while increases stability and the range of stability for higher enrichment values compared to an open ecosystem with no recycling (Fig. 3.2). Nutrients recycled by the autotrophs, on the other hand, strictly decrease stability compared to the open ecosystem with no recycling (Fig. 3.2). Recycling by both biotic compartments results in effects similar to that of consumer recycling with greater resilience (Fig. 3.2).

In order to simplify the analysis of the full meta-ecosystem, we assume both ecosystems have identical parameter values and we use parameter values that guarantee a stable equilibrium in each ecosystem when there is no nutrient diffusion. The equilibrium that both ecosystems achieve is called a spatially homogeneous (flat) solution and we analyze its linear stability using the analytical techniques developed by [86]. To determine the linear stability of the flat solution, one must determine the eigenvalues of a series of k matrices, $V(k)$, and all the eigenvalues of each matrix $V(k)$ must have negative real parts for the spatially homogeneous solution to be stable (see Chap. 4 for details). The formula for each $V(k)$ matrix is

$$\mathbf{V}(k) = \mathbf{J} + \lambda_k \mathbf{M} \tag{3.7}$$

where $\mathbf{J}$ is the Jacobian matrix of the well-mixed ecosystem evaluated at the flat solution, $\mathbf{M}$ is the diffusion coefficient matrix and λ_k are the eigenvalues of the connectivity matrix $\mathbf{C}$, which describes the spatial arrangement of the ecosystems. For our study, $\mathbf{M}$, $\mathbf{C}$ and λ_k are

$$\mathbf{M} = \begin{bmatrix} d_N & 0 & 0 \\ 0 & d_A & 0 \\ 0 & 0 & d_C \end{bmatrix}, \mathbf{C} = \begin{bmatrix} -1 & 1 \\ 1 & -1 \end{bmatrix}, \lambda_1 = 0, \lambda_2 = -2 \tag{3.8}$$

The flat solution can also be a limit cycle or a chaotic attractor, but we only investigate the stability of limit cycles numerically.

Using Eq. 3.7 and the Routh-Hurwitz conditions to determine the stability of the flat solution (an equilibrium), we discover that there exists a positive value of d_N, denoted as $d_{N,crit}$, at which point the flat solution is no longer stable. This result is not sensitive to the values of model parameters and occurs for both the closed and open meta-ecosystem, though the value of $d_{N,crit}$ is sensitive to parameter values. Numerical examination of the eigenvalues of $\mathbf{V}(i)$ matrices indicates that at $d_{N,crit}$, the system undergoes a Hopf bifurcation, or rather a spatial Hopf bifurcation as it is caused by diffusion.

To further understand the model beyond the stability analyses, the dynamical regimes and the regional stability of meta-ecosystems can be examined numerically using bifurcation plots (local extrema of time series) and recording the minimum abundance of all trophic levels as we vary d_N. We can use covariance as a statistical measure spatial synchrony between ecosystems ($N_1 - N_2$, $A_1 - A_2$ and $C_1 - C_2$) without attempting at this stage to evaluate phase synchrony (see Sect. 3.4 below). Covariance is also used to quantify net coupling between net autotroph growth and both nutrient diffusion and recycling, in relation to diffusion rate d_N.

Numerical results revealed similar dynamical behavior of spatial covariance and regional stability to increasing d_N for the meta-ecosystems whether they were closed or open (with no contribution from recycling). For example, in a closed meta-ecosystem, increasing d_N above its critical level destabilizes the stable equilibrium in the local ecosystems and leads to stable oscillatory dynamics with 2 local extrema (for consumers and autotrophs) or 4 local extrema (nutrients) to emerge (Fig. 3.3a).

At the same critical d_N required for destabilization of local dynamics, we witness negative spatial (between ecosystem) covariance across all trophic levels in the coupled-ecosystems Fig. 3.3b. As the coupling is further increased, the spatial covariance of nutrients becomes positive and phase synchronous while the spatial covariance of the autotrophs and the spatial covariance of the consumers remain negative, indicating a nutrient-induced anti-phase synchrony of biotic levels Fig. 3.3b.

Greater understanding of spatiotemporal instabilities above the critical d_N value can be gained from the analysis of feedbacks between nutrient flow and population growth across trophic levels as measured by their covariance. There is a positive covariance between net autotroph growth and nutrient diffusion when spatiotemporal instabilities occur (Fig. 3.3c). In contrast, nutrient recycling co-vary negatively with autotroph growth (Fig. 3.3d). The above results indicate a positive feedback (positive covariance) induced by the spatial process of nutrient diffusion, while recycling (or nutrient input) is linked to a negative feedback (negative covariance). An increase pagebreakin autotroph growth in one ecosystem depletes nutrients in that

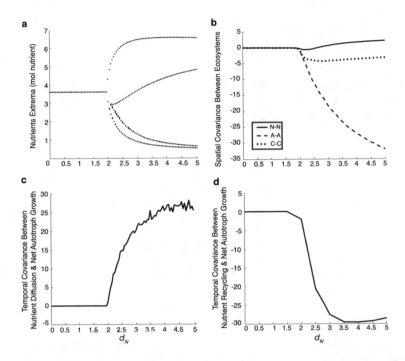

Fig. 3.3 Analysis of the effect of increasing the diffusion coefficient (dN) in a closed meta-ecosystem on: **a** local stability, **b** spatial synchrony between ecosystems, **c** the covariance between nutrient diffusion and autotroph growth and **d** the covariance between nutrient recycling and autotroph growth. Adapted from [115]

ecosystem, allowing nutrient input from the second ecosystem through diffusion, which positively feeds back on autotroph growth (e.g. Fig. 3.4).

These results suggest that the bifurcation to spatiotemporal complexity is solely diffusion driven. However, recycling strongly affects the strength of feedbacks - positive and negative—between nutrient availability and growth that in turn drive the critical diffusion rate leading to the spatial Hopf bifurcation (Fig. 3.5). The specific effect of recycling on the diffusive instability depends on enrichment. Meta-ecosystems with no recycling are more resistant to diffusion-induced instabilities when exposed to weak enrichment ($4 < S^* < 7$) compared to strong enrichment ($S > 8$; Fig. 3.5). The open meta-ecosystem with perfect recycling is always less stable than the closed meta-ecosystem due to the loss of nutrients (Fig. 3.5). In all cases, greater enrichment results in lower $d_{N,crit}$ values, so the gains in stability are relative (Fig. 3.5).

While spatiotemporal stability is lost through the positive feedback between nutrient diffusion and autotroph growth, nutrient recycling controls the strength of this nutrient-growth feedback and determines the critical diffusion rates at which diffusion-driven instabilities occur. Nonlinear meta-ecosystem dynamics is able to

Fig. 3.4 The dynamics of a closed meta-ecosystem after they are perturbed from an unstable flat solution. The colors of the lines indicate the ecosystem, while the solid lines, the thick dashed lines and thin dashed lines denote nutrients, autotrophs and consumers, respectively. Adapted from [115]

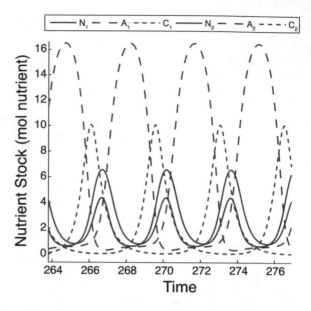

Fig. 3.5 Comparing $d_{N,crit}$ for a closed (black line), open with perfect recycling (grey line), open with imperfect recycling (dashed dotted grey line) and open with no recycling (dashed grey line) meta-ecosystems as a function of nutrient enrichment S^*. Adapted from [115]

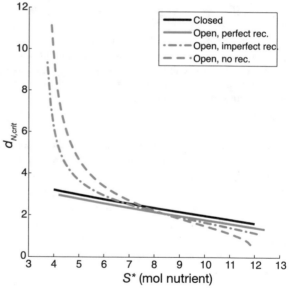

destabilize the local dynamics of the biota in each ecosystem at levels of total nutrients that could not induce such destabilization in a well-mixed ecosystem. Nutrient
recycling has been evoked as a mechanism of either stabilization or destabilization
of ecosystem dynamics [106]. Nonlinear meta-ecosystem dynamics indicate that
recycling has an ambiguous relationship with ecosystem stability. This relationship
is highly dependent on the amount of enrichment within the ecosystem and on the
source of the nutrients.

3.3.1 Spatial Dynamics with Multiple Ecosystem Compartments

We saw in Chap. 2 that the retention of organic matter in a detritus compartment
can have important implications for species coexistence and community structure
in meta-ecosystems (see Sect. 2.3). That same compartmentalization of organic and
inorganic matter into separate nutrient and detritus stocks can also have important
implications for the stability of nonlinear meta-ecosystem dynamics, including their
response to enrichment. Reference [54] used the 2-patch meta-ecosystem model
from equations (2.7a) and replaced linear functional responses f used by [57] with
nonlinear Holling Type II saturating functional responses (see Eq. 3.1 and Fig. 3.6).

   We can first look at the effect of a single spatial flow (d_N, d_P, d_C, or d_D) while
setting the others to 0. In this case, linear stability analysis of equilibrium points
show that increasing spatial flows of nutrient and detritus compartments can have a
stabilizing effect (help maintain stable equilibrium) in response to spatially heterogeneous enrichment and destabilizing effect (promote oscillatory dynamics) when
enrichment is homogeneous across ecosystems. In other words, In a homogeneous
environment (i.e. ecosystems with similar local fertility; $I_1 = I_2$), ecosystems are
less robust to enrichment when they are connected by spatial flows of detritus or
nutrient, than isolated ecosystems (Fig. 3.7, dashed lines).

   Another consequence of introducing nonlinear functional responses in meta-
ecosystems with separate organic and inorganic stocks is the emergence of multiple stable states. Consumer diffusion can for example generate multiple equilibria
when intermediate diffusion rates are combined with high values of regional fertility (Fig. 3.8). Two nontrivial equilibria are then observed: oscillatory dynamics
in complete phase synchrony (Fig. 3.8c, e), or an asymmetric source-sink structure
(Fig. 3.8d, f, and orange line in b). The stable asymmetric equilibrium observed under
these conditions reproduces the source-sink behavior described in Chap. 2 with linear
functional responses (Sect. 2.3): an initial perturbation allows the producer of one of
the two ecosystems to exploit its abundant resource. This ecosystem produces numerous consumers and therefore becomes a net exporter. The consumers are exported to
the second ecosystem, where they prevent the growth of the producer despite abundant inorganic resources. Subsequently, the organic matter brought by the consumers
is mainly stored in the nutrient compartment of the second ecosystem, which becomes

a net importer. This results in a stabilizing spatial asymmetry in ecosystem control. Hence, consumer spatial flows allow a regional stabilization for a set of intermediate diffusion rates, even when enrichment reaches high levels. This increasing stability can be interpreted as an emerging spatial heterogeneity in the direction of trophic regulation (top-down vs. bottom-up controlled ecosystems). However, reference to directional trophic regulation implied when using terms like top-down and bottom-up has limited relevance for understanding food-chain and food web dynamics once recycling is considered explicitly. Recycling introduces circularity in trophic control that can be better understood from the description of full regulation pathways across ecosystem compartment that can include combinations of top-down and bottom-up effects [99].

Results from [54] tell us that the nature of the compartment diffusing between ecosystems determines whether diffusion enhances enrichment-induced instabilities: Spatial flows of nonliving compartments (nutrients or detritus) are destabilizing, and intermediate spatial flows of consumers can switch dynamics from stable oscillations to stable equilibrium even under high enrichment.

Taken together, the relatively simple 2-patch meta-ecosystems that have been studied so far show the complex interaction between recycling and nonlinear trophic interactions in a spatial context. The linear recycling functions assumed in [115] and in [54] promote the destabilizing effects of nonlinear trophic interactions when

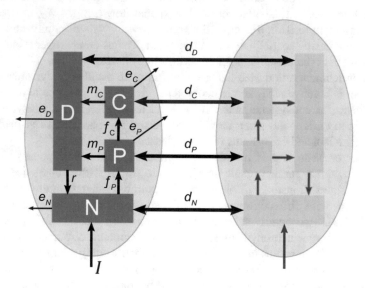

Fig. 3.6 Metaecosystem model from [54]. In each of the two ecosystems, the primary producer P consumes the inorganic nutrient N and is consumed by the primary consumer C. P and C produce detritus D at respective rates m_P and m_C, mineralized into N at a rate r. The functional responses of the organisms, f_P and f_C, take a Holling type II form. N receives constant input I from the outside. Each compartment loses material at constant output rates e_N, e_P, e_C, e_D. The ecosystems are connected by spatial flows between their homologous compartments N, P, C, D, according to constant diffusion rates $d_N, d_P, d_C,$ and d_D, respectively

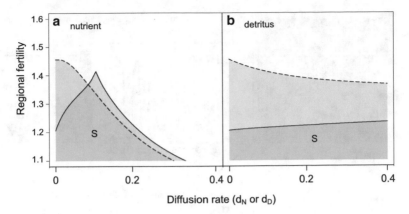

Fig. 3.7 Stability with a single spatial flow. Each panel represents the stability isoclines under changes of one diffusion rate (the others set to 0)—either **a** d_N or **b** d_D—versus regional fertility of the metaecosystem ($(I_1 + I_2)/2$). The stability isoclines are the pairs of parameter values for which the dominant eigenvalue equals 0 either in a homogeneous meta-ecosystem with $\Delta I = 0$ (dashed lines) or in a heterogeneous meta-ecosystem with $\Delta I = 0.5$ (solid lines). Stability isoclines delimit the gray parameter space (S) where the dominant eigenvalue is negative, and thereby the equilibrium is stable. Adapted from [54])

recycling is accompanied by spatial fluxes of non-living compartments. The onset of oscillatory dynamics is often interpreted as an important predictor of local persistence, which is inversely correlated with the amplitude of population fluctuations (e.g. [123]). Oscillations also play a role in species coexistence by creating species-specific variability on which niche differentiation can operate [5], including via non-linear averaging (Jensens's inequality effect, [155]). The fact that spatial fluxes of matter can play such an important role to predict oscillatory dynamics and even multiple stable states in response to enrichment means that nonlinear meta-ecosystems can provide a minimal ecological system to integrate the study a broad range of mechanisms for the maintenance of variability in population abundance.

Understanding non-equilibrium dynamics in meta-ecosystems lets us integrate the cycling of organic and inorganic matter to the large body of literature on the role of variability, both spatial and temporal, for key ecological properties such as persistence, coexistence, and productivity. The role of spatial fluxes also forces us to consider the spatial relationship between local oscillators and their scaling-up to meta-ecosystems: strong local oscillations can be compatible, and even promote regional population persistence if these local oscillations help maintaining spatial heterogeneity among ecosystems. While a number of statistical approaches can be adopted to characterize relationships between temporal and spatial variations across scales, they all relate, in some way at least, to the problem of spatial synchrony. As we will now see, spatial synchrony can be studied both as a statistical property of spatiotemporal series, and as a mathematical formalism to study spatial dynamics. As we review studies of spatial synchrony as an important mechanisms explaining

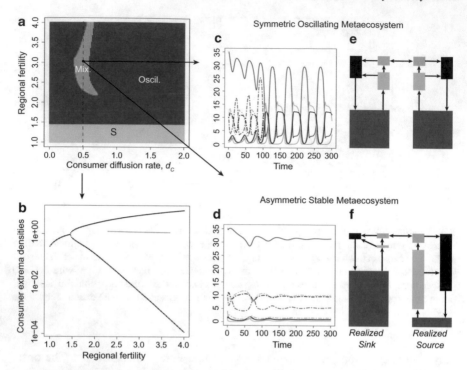

Fig. 3.8 Multiple equilibria with intermediate consumer diffusion rates and high enrichment levels in homogeneous meta-ecosystems. **a** Areas of stability for an extended parameter space (d_C from 0 to 2 and regional fertility from 1 to 4) for the homogeneous case, with $\Delta I = 0$. Gray area (S), stable equilibria (negative dominant eigenvalue). Red area (Oscil.), unstable equilibria (positive dominant eigenvalue). Orange and purple areas have multiple equilibria, either all unstable (purple area) or one unstable and two stable (Mix.; orange area). **b** Bifurcation diagram of stable states showing consumer extreme densities (spatial average) according to regional fertility for the consumer diffusion rate $d_C = 0.5$ (dashed line in a). **c–f** illustrate the two equilibria types for the pair of parameters ($d_C = 0.5$, regional fertility = 3). **c** and **d** show the dynamics of all the compartments. **e** and **f** show the relative densities of the different compartments (heights are proportional to the temporal mean density at equilibrium). Adapted from [54]

nonlinear metapopulation and metacommunity dynamics, we will also cover its more recent integration to meta-ecosystem theory.

3.4   Spatiotemporal Heterogeneity and Spatial Synchrony

The potential complexity of spatial dynamics in meta-ecosystems can be distilled down to one property: deviation from spatial (among location) in-phase synchrony of local temporal fluctuations. This property is important because it is a necessary and

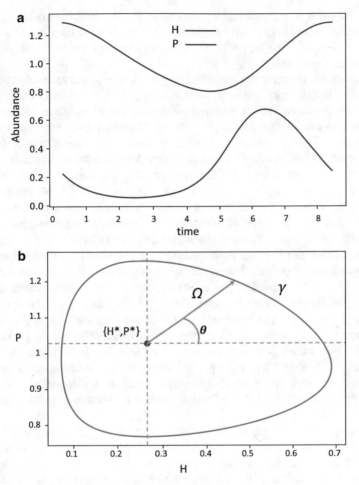

Fig. 3.9 Examples of time series of predator (P) and prey (H) abundances from the Rosenzweig-MacArthur model (Eq. 3.1), **a** and of corresponding phase plane trajectory **b** illustrating the phase angle θ and the amplitude Ω along the solution orbit γ measured from the unstable prey and predator equilibrium H^*, P^*

sufficient condition for spatiotemporal heterogeneity. However, it requires a formal definition in order to be operational as a metric of spatial dynamics. In ecology, synchrony has been associated with coherent variations in population density [104]. It can be defined as a statistical property and measured as statistical correlation or coherence [1, 12, 93].

Synchrony can have a more specific definition when applied to time series containing periodic components. In such cases, a periodic time series (Fig. 3.9a) can be represented by its phase and amplitude (Fig. 3.9b). The phase is the 2π-periodic angular position of the variable along a cycle, and the amplitude measures the distance of the cycle from a reference point, typically the unstable equilibrium. In a 2-dimensional system such as the Rosenzweig-MacArthur model (Eq. 3.1), predator-prey cycles (Fig. 3.9a) can be easily visualized as an orbit from which the phase and amplitude can be measured (Fig. 3.9b).

In addition to extracting the phase and amplitude from time series, dynamic systems can similarly be rewritten with phase replacing abundances as the state variable. Synchrony is then measured as the phase difference among locations or species [12, 13]. Phase synchrony, including in-phase synchrony, is typically associated with the existence of an equilibrium zero phase difference among time series (Fig. 3.10a). Statistical and phase-based definitions are not equivalent because statistical correlation provides limited information on the degree of phase synchrony. As a result, deviation from in-phase synchrony, often referred to as asynchrony, cannot be unequivocally detected from loss of correlation or of correspondence between phases. For random variables, the term asynchrony is used to describe statistical independence or even negative correlation. For periodic signals, deviation from in-phase synchrony is equivalent to loss of spatial homogeneity and provides no additional information on dynamical properties of phase differences among locations. Phase synchrony can be more broadly viewed as phase locking defined as the existence of an equilibrium phase difference between time series. We can see that synchrony is really defined as a particular dynamical state, in this case an equilibrium, of phase differences. Synchrony leads to spatial homogeneity of frequencies among periodic signals (Fig. 3.10b) rather than of abundance. From this definition, phase asynchrony can be defined as a deviation from phase locking, and as a non-equilibrium phase difference dynamics associated with spatially-heterogeneous frequencies. That definition of phase asynchrony leaves of course room for a number of patterns of spatial dynamics.

As for any dynamical system, the loss of stability of the local phase-synchronous equilibrium between 2 locally oscillating systems can lead to a number of dynamical regimes. One form of phase asynchrony can arise from spatial heterogeneity when communities with different natural frequencies of local oscillations are (weakly) coupled and the heterogeneous frequencies persist (Fig. 3.10c, d). This asynchrony can be synchronized by (sufficiently strong) dispersal [13, 93], or by a common environmental forcing [19]. A different form of phase asynchrony can arise when spatially heterogeneous environmental fluctuations modulate the frequencies of local population fluctuations [3] (Fig. 3.10e, f). Frequency modulation can more generally emerge from weak symmetric [6, 62] and asymmetric [130] coupling, without any fluctuation or heterogeneity in the environment.

We can first study spatial synchrony and asynchrony as equilibrium and non-equilibrium phase dynamics in networks of weakly-coupled oscillators [162]. In ecology, this formalism has first been applied to trophic interactions leading to local fluctuations of abundance [8, 50]. As we will see next, an important advantage of assuming weak coupling is the simplification of dynamic systems with multiple state variables of abundance being reduced to a single phase variable (H and P vs θ in Fig. 3.9b). This simplification is possible because by assuming weak coupling, we can safely ignore any impact of coupling on the shape of local orbits (γ in Fig. 3.9b)

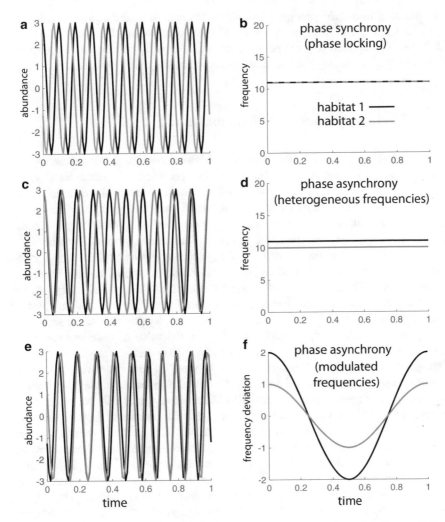

Fig. 3.10 Examples of time series of abundance (**a**, **c**, **e**) and of corresponding frequency (**b**, **d**, **f**) corresponding to phase synchrony (**a**, **b**) and to phase asynchrony driven by either spatially heterogeneous habitats (**c**, **d**) or by temporal frequency modulation (**e**, **f**). Adapted from [62]

and thus on amplitude (Ω in Fig. 3.9b). In ecology, phase dynamics has mostly been studied in metapopulations and metacommunities, and no study of meta-ecosystems that the authors are aware of has applied this formalism. However, a number of nonlinear meta-ecosystem theories have analysed spatial synchrony of phase and amplitude emerging from the movement and cycling of (in)organic matter. These studies have so far stayed away from the assumption of weak coupling, and are thus predicting how asynchrony of both phase and amplitude respond to strong coupling that characterizes many natural ecosystems. We believe the dynamics of phase and amplitude offer promising avenues to study the maintenance and ecological implications of spatiotemporal variability in meta-ecosystems. Because these tools still have to be applied to meta-ecosystems, the following section can be skipped without loss of continuity with material presented in the next chapters.

3.4.1 Phase Dynamics in a Metacommunity

In its simplest form, a system of coupled oscillators consists of two discrete communities, each with one predator and one prey species. Their dynamics is characterized by nonlinear feedbacks, which allows for limit-cycle fluctuations. Coupled oscillators are defined by parameters of local dynamics and of spatial coupling between communities. As we saw earlier, they can lead to spatiotemporal heterogeneity in the abundance of predators and preys that are not perfectly synchronous across patches. This deviation from in-phase synchrony can be explained by heterogeneous environmental (external) forcing, by differences (spatial heterogeneity) in the parameters affecting the period of local limit cycles (proper frequency), or by limited movement of biomass among communities. One way to predict the onset of spatially-heterogeneous dynamics is to conduct a stability analysis of the spatially-homogeneous solution [84, no spatial variation in the value of state variables; see also Sect. 3.3]. This method predicts the onset of spatial dynamics for arbitrary values of spatial fluxes between communities, but it is based on small perturbations of the spatially-homogeneous solution, and provides little information on the properties of resulting spatiotemporal heterogeneity [85]. Alternatively, it is possible to simplify the description of the full dynamics to that of a phase difference in the periodic fluctuations of predators and preys between communities. This method assumes periodic oscillations and its analysis is based on weakly coupled communities. However, when such restrictive conditions apply, the stability analysis of equilibrium phase differences predicts a great diversity of spatiotemporal regimes that are characterized by phase locking, including the existence of multiple stable heterogeneous states [50]. We can also observe non-equilibrium dynamics of phase differences with associated periodic and spatially-heterogeneous fluctuations in the frequencies of oscillations in population abundance [62].

We now review the method allowing for the stability analysis of phase difference between 2 spatially distinct predator-prey communities coupled by spatial fluxes of both predator and prey individuals. This method consists in rewriting the equations of predator and prey dynamics as a single equation of phase dynamics in each patch. We can then write the phase dynamics in each patch as a single equation of phase difference dynamics that we can analyse for local stability. In order to reach that point, we will need to assess the effect of spatial fluxes on phase difference. Reader can skip the following derivations and go directly to the discussion paragraphs below. However, the mathematical formulation and analysis of weakly-coupled oscillators is useful to highlight the power of simplification of this approach and its assumptions that are both important to guide to its future application to meta-ecosystems.

Following [50], we first use a non-dimensional version of the two-patch model (3.3) for prey (h_i) and predator (p_i) density on patch $i, j = 1, 2$ ($j \neq i$):

$$\frac{dh_i}{d\tau} = \frac{1}{\epsilon}\left(h_i(1 - kh_i) - \frac{h_i p_i}{1 + h_i}\right) + d_h(h_j - h_i) \qquad (3.9a)$$

$$\frac{dp_i}{d\tau} = \frac{h_i p_i}{1 + h_i} - \eta p_i + d_p(p_j - p_i) \qquad (3.9b)$$

with the following substitution in system (3.3):

$h_i = (1/b)H_i$, $p_i = [a/(rb)]C_i$, $\tau = at$, $k = b/K$, $\eta = m/a$, $\epsilon = a/r$, $d_h = D_H/a$, $d_p = D_C/a$.

The non-dimensional version of the Rosenzweig-MacArthur model reduces the number of parameters compared to the original model, and let us focus on non-dimensional parameters assessing the temporal scale of prey vs predator growth (see below). We then cast model system (3.9) into vector form for $X_i = (h_i, p_i)^T$ as

$$\frac{dX_1}{dt} = F(X_1) + \delta W(X_1, X_2), \qquad (3.10a)$$

$$\frac{dX_2}{dt} = F(X_2) + \delta W(X_2, X_1), \qquad (3.10b)$$

with

$$F(X_i) = \left(\frac{1}{\epsilon}\left(h_i(1 - kh_i) - \frac{h_i p_i}{1 + h_i}\right), \frac{h_i p_i}{1 + h_i} - \eta p_i\right), \qquad (3.11a)$$

$$W(X_i, X_j) = \left(\frac{d_h}{\delta}(h_j - h_i), \frac{d_p}{\delta}(p_j - p_i)\right) \qquad (3.11b)$$

where $\delta = max(d_h, d_p)$ is assumed small (weak spatial coupling). We can now shift our point of view to study spatial synchrony. To that end, we assume that the system is oscillating, and we describe only the dynamics of the phase of the system along the periodic orbit on each patch.

We denote $\theta_i \in [0, 2\pi]$ as the phase variable of model (3.10) along the periodic orbit $\gamma(t)$ on patch i. Following the theory for weakly connected networks, [79] proved that these phase variables satisfy approximately the following two-dimensional model

$$\frac{d\theta_1(t)}{dt} = \Omega + \delta H(\theta_2 - \theta_1) \qquad (3.12a)$$

$$\frac{d\theta_2(t)}{dt} = \Omega + \delta H(\theta_1 - \theta_2). \qquad (3.12b)$$

The 2π-periodic function H is given by

$$H(x) = \frac{1}{T} \int_0^T \hat{\gamma}(t) \cdot W(\gamma(t), \gamma(t + x/\Omega)) dt, \tag{3.13}$$

where $\hat{\gamma}(t)$ is the unique solution of

$$\frac{d\hat{\gamma}(t)}{dt} = -DF(\gamma(t))^T \hat{\gamma}(t) \tag{3.14}$$

with the normalization condition $\hat{\gamma}(t) \cdot \gamma'(t) = 1$.

Next, we introduce the phase deviation variables ϕ_i through $\theta_i(t) = \Omega t + \phi_i(t)$. System (3.12) can now be rewritten as

$$\frac{d\phi_i(t)}{dt} = H(\phi_j(t) - \phi_i(t)), \tag{3.15}$$

for $i, j = 1, 2$ and $j \neq i$. Then we can write the single equation for the phase difference $\phi := \phi_1 - \phi_2$ as

$$\frac{d\phi(t)}{dt} = G(\phi) := \delta(H(-\phi) - H(\phi)). \tag{3.16}$$

A constant solution ϕ^* satisfying $G(\phi^*) = 0$ represents the phenomenon of phase-locking. We say that the system is in-phase synchronized when it is locked at $\phi^* = 0$, and that the system is anti-phase synchronized when it is locked at $\phi^* = \pi$. The sign of $G'(\phi^*)$ determines the local stability of the phase-locked state ϕ^* in model (3.16). If it is positive (negative), the steady state ϕ^* is unstable (stable). The absolute value $|G'(\phi^*)|$ indicates the rate of convergence to phase locking. Using this framework, [50] showed that phase locking of weakly coupled predator-prey communities is determined by the separation of temporal scales between predator and prey dynamics (Fig. 3.11). Decreasing any of ϵ, k or η increases the difference in time scales between the two species. A small ϵ, $(0 < \epsilon < 1)$ increases the speed of every aspect of the intrinsic prey rates relative to those of the predator and in the limit as ϵ goes to zero, the system becomes a relaxation-like oscillator characterized by pulses of predator growth. Decreasing k increases the carrying capacity, thereby increasing both the magnitude of the prey outbreak and the time between outbreaks. A slower predator death rate η increases the amount of time necessary for the predator population to become sufficiently small so as to allow for a prey outbreak, increasing the amount of time in the cycle with low prey population.

When prey and predators have similar characteristic time scales, the result is a synchronous stable steady state and an unstable anti-synchronous steady state. Increasing the separation of these time scales yields both stable in-phase and out-of-phase (often anti-synchronous) synchronous steady states with an additional unstable phase-locked synchronous steady state between the two stable states (Fig. 3.11).

Phase dynamics greatly simplifies the description of coupled systems by reducing the explicit dynamics of abundance of each species at each location to the dynamics

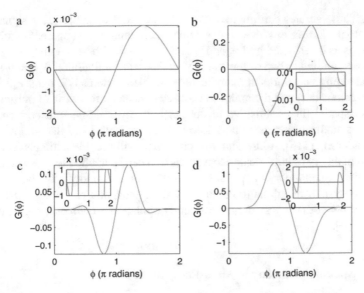

Fig. 3.11 Plots of the G-function for different level of time scale separation between predator and prey growth controlled by parameters ϵ, k, and η. Insets show behavior not apparent because of the large magnitude of in the full figures. The G-function describes the rate of change of phase between two patches. The steady states or phase locked states occur when G = 0, and are stable when the slope of G is negative at the fixed point and unstable when the slope is positive. The larger maximum magnitude of the phase response curve corresponds to a larger maximum magnitude of these G-functions and faster convergence to the stable phase locked states and is the result of a greater separation in time scales. Case **a** with $\epsilon = 0.1$, $k = 0.4$, and $\eta = 0.4$ will have the slowest convergence of the four while cases (**b**; $\epsilon = 0.05$, $k = 0.28$, and $\eta = 0.38$), (**c**; $\epsilon = 0.1$, $k = 0.3$, and $\eta = 0.3$), and (**d**; $\epsilon = 0.1$, $k = 0.4$, and $\eta = 0.15$) will all have very fast convergence at certain times in the phase. Adapted from [50]

of their phase difference between locations. The stability analysis of phase differences further limits their complexity by imposing weak spatial coupling. With such simplifications come some limits to the applicability of phase analysis to natural systems. Methods presented above assume linear phase evolution and weak coupling, together allowing the description of dynamical systems from the perspective of their phase with no consideration of amplitude. These are not absolute limitations and they challenge us to expand the current toolbox as it is applied in ecology. Statistical approaches based on the extraction of both phase and amplitude from time series in trophic metacommunities have shown how amplitudes of multiple time series can remain asynchronous even when their phases are locked in synchrony [13]. The complementary information provided phases and amplitudes provides a powerful signature of spatial dynamics, and allows predicting the path from phase locking to full correlation (phase and amplitude synchrony) between time series [13]. It also stresses the importance of studying feedbacks between phase difference and amplitude variations when spatial coupling cannot be assumed weak, which can lead to a much broader range of spatiotemporal phenomena that are not captured by phase locking. These studies can improve the use of spatial dynamics as a signature of local ecological processes, even for systems where there is no apparent regularity in fluctuations as determined by frequency and amplitude. Phase stability analysis has mostly been limited to two or few coupled ecological oscillators [62] and it still has to be scaled up to networks of many locally-coupled systems [77, but see].

Applications of phase dynamics to ecology lacks an integration of nutrient recycling and spatial fluxes. Positive feedbacks between nutrient transport and recycling can actually turn connectivity into a driver of both local fluctuations and of their asynchrony [115]. Addressing this challenge will provide a more general non-equilibrium theory of large-scale ecosystems that can be applied to ecosystem-based management. Such theory of spatial synchrony is based on local positive feedbacks linking ecosystem functions to community dynamics, and can predict patterns of covariance between ecosystem functions and population growth over regional scales [11, 52, 148].

3.4.1.1   Spatial Synchrony in Meta-Ecosystems

There are numerous studies of spatial synchrony in metacommunities, but very few on nonlinear meta-ecosystems. The cycling and movement of (in)organic matter has been studied in linear meta-ecosystems displaying equilibrium dynamics, and recent studies have explored the specific role of nutrients for the maintenance of both local (within-ecosystem) and regional (among-ecosystem) fluctuations and spatial heterogeneity that can be captured by spatial synchrony. None of these studies have been conducted by reducing full models to phase dynamics under the strict assumption of weak coupling between local ecosystems. As a result, our understanding of nutrient and detritus movement as a driver of spatial (a)synchrony includes variations in both phase and amplitudes and is mostly based on numerical simulations. One important message conveyed by these studies is that nonlinear feedbacks introduced by nutrient and detritus dynamics can make spatial heterogeneity in meta-ecosystems very robust to in-phase and amplitude synchronization that would be expected as spatial fluxes are increased, thus violating assumptions of weakly-coupled oscillator theories.

For example, [115] showed that increasing the rate of nutrient movement in system Eq. 3.4 leads to both the loss of local stability of the local equilibrium (Fig. 3.3a) and to out-of-phase spatial synchrony of emerging oscillatory dynamics (Fig. 3.3b). Spatial in-phase synchrony of primary producers and consumers can be maintained under high nutrient diffusion rate, resulting in the doubling of frequency of in-phase nutrient oscillations with heterogeneous amplitudes (Fig. 3.4).

Most studies of spatial synchrony in meta-ecosystems were conducted with added levels of ecological complexity and will be covered into more details in Chap. 4. Here we simply discuss how feedbacks introduced by nutrient and detritus compartments can affect the spatial synchrony of phase and amplitudes.

In larger and more complex meta-ecosystem networks [116, see Chap. 4 for details], spatial flows of nutrients can also have an important impact on the mainte-

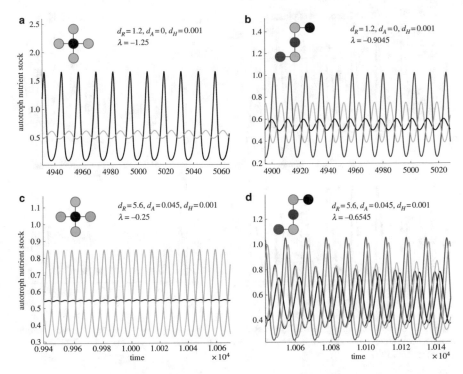

Fig. 3.12   The effects of meta-ecosystem network structure on local ecosystem dynamics after the local equilibrium solution is destabilized by high nutrient movement. The colours of the graphic insets indicate which time series is to be found in the local ecosystem, which means if two local ecosystems share the same colour, they have the same temporal dynamics. With d_A equal to zero, the two meta-ecosystem configurations (**a**, **b**) show large oscillations in their most connected ecosystems, though the patterns differ greatly between them. With d_A equal to 0.045, meta-ecosystems (**c**, **d**) continue to show oscillations, but the most centralized ecosystems no longer show the highest amplitude oscillations; however, the temporal and spatial dynamics differ greatly between them. Adapted from [116]

nance of heterogeneity in both phase and amplitude of oscillations. While the stability analysis at the network level is presented in Chap. 4, the level of nutrient movement contributes extending both the magnitude and spatial scale of the perturbation it propagates across the network. By doing so, it breaks the network-level symmetry of spatial synchrony: In ecosystems connected only by relatively weak nutrient fluxes, ecosystems sharing the same level of connectivity (number of links) oscillate in full (phase and amplitude) synchrony while heterogeneous phases and amplitudes are maintained among ecosystems with varying number of links (Fig. 3.12a, b). Increasing the rate of nutrient and autotroph movement breaks that symmetry and results in phase and amplitude differences between each pair of ecosystems, leaving no recognizable pattern at the network level (Fig. 3.12d).

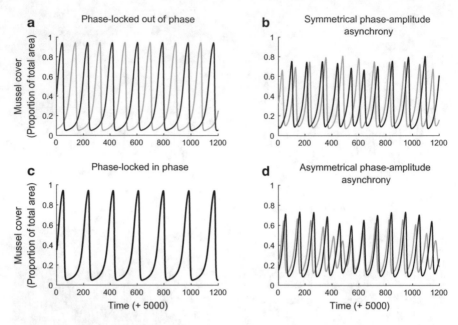

Fig. 3.13 (**a–d**) Local mussel cover time series under possible spatial synchrony regimes, colours denote ecosystem 1 (black) and 2 (yellow). From **b** to **e**, d_M values are 0, 0.16, 0.23 and 0.34. Adapted from [87]

Spatial synchrony is also strongly affected by the movement of detritus material, which can be in part driven by the non-resource impact of detritus on local ecosystem dynamics. As we reviewed above in Sect. 3.4.1, pulse-relaxation type oscillations that are characterized by non-sinusoidal and abrupt changes, are prone to the maintenance of non-zero phase locking. Reference [87] showed that similar pulse-relaxation oscillations could be driven by the effect of accumulating detritus in mussel bed ecosystems on the structural stability of those beds facing wave disturbances. Such non-resource effect of the detritus compartment, detailed in Chap. 4, introduces a positive and nonlinear feedback between the growth of mussel beds and their fast (sudden) disturbance by waves, resulting in local cycles of slow growth and fast disturbance of mussel cover. These local pulse-relaxation oscillations, when coupled through larval dispersal of mussels between two ecosystems display spatially-heterogeneous oscillations that are very resistant to increasing dispersal that would be expected to lead to in-phase synchrony. As mussel dispersal rate is increased (Fig. 3.13 from a–d), spatial dynamics display multiple transitions synchronous to asynchronous dynamics, going from out-of-phase synchrony (Fig. 3.13a) to symmetrical phase-amplitude asynchrony, meaning that both ecosystems show similar long-term dynamics (Fig. 3.13b). Further increasing mussel dispersal locks the system be to in-phase synchrony (Fig. 3.13c), which could be interpreted as the expected behavior of the system in the limit where high dispersal rates lead to perfect mixing between ecosystems. But instead, even higher values of dispersal lead to a new

Fig. 3.14 Relationship between the amplitude and the frequency of local mussel cover dynamics. Colours indicate the strength of larval dispersal (d_M) from low (black) to high (yellow). Adapted from [87]

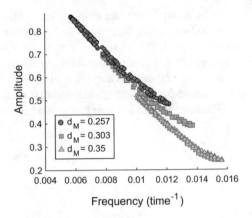

asynchronous regimes of both phases and amplitudes with long-term heterogeneity, or asymmetry, between ecosystems where one ecosystem displays consistently stronger amplitudes (Fig. 3.13d). In all cases where phase-amplitude asynchrony is observed, the fluctuations in frequency and amplitude are bounded, periodic, and negatively correlated (Fig. 3.14). In other words, frequency and amplitude are dynamically coupled. Given that both frequency and amplitude have important ecological implications for stability and persistence, understanding their coupling becomes an important challenge.

These results offer an opportunity to integrate meta-ecosystem theories into the broader study of pulse dynamics over networks applied to epidemics [74], pest outbreaks [175], and neural networks [83]. Over regional scales, our results show how pulsed disturbance cycles can maintain spatial phase-amplitude asynchrony among strongly coupled ecosystems characterized by long-term fluctuations in the frequency and amplitude of abrupt changes. In contrast, regular oscillations lock in-phase under much weaker spatial fluxes, at which point dispersal and spatial fluxes of detritus have no regional effects. Fluctuating asynchrony has been observed in many long-term time series of population abundance [20], and is typically attributed to long-term environmental fluctuations [76]. On the other hand, aggregation processes can explain the onset and spatial synchrony of abrupt changes in forest ecosystems citebib146, and the results reviewed above further show how fluctuating phase and amplitude synchrony can emerge from strong coupling among ecosystems driven by foundation species such as trees, salt marshes and bed-forming mussels.

In this chapter we have centered our attention on nonlinearities that characterize many ecological interactions and individual responses to the environment, and on non-equilibrium dynamics they predict. We saw that the movement and cycling of matter, even when described by linear functions can interact with nonlinear species interaction and affect the maintenance of local oscillations and of their spatial heterogeneity (or synchrony) across whole meta-ecosystems. We will now see that nonlinear and resulting non-equilibrium dynamics maintain a central role to extend meta-ecosystem theories in order to make them more applicable to natural ecosystems, or at least to create stronger connections with other ecological theories such as ecological stoichiometry, ecological networks, food webs and foundation species. By integrating species interactions to the cycling and spatial fluxes of matter, one could argue that the meta-ecosystem framework provides an appropriate level of abstraction to integrate and even generalize these and other ecological theories. In any case, whenever ecological theories increase their relevance to natural ecosystems, nonlinear dynamics is often just around the corner.

Chapter 4
Diversity of Meta-Ecosystems: Spatial Topologies, Species Diversity, Stoichiometries and Non-trophic Fluxes

Abstract The application of meta-ecosystem theories to natural systems requires extending simple models to account for complex spatial networks, species diversity, and for multiple limiting nutrients. We first show that predictions from 2-patch meta-ecosystem models don't necessarily scale-up to larger spatial networks. The expansion of single ecosystem compartments at a given trophic level also allows for the explicit evaluation of the effects of biodiversity on ecosystem functioning. In addition, we can include multiple limiting nutrients to examine the impact of the coupling of biogeochemical processes within ecosystems that occur across different timescales due to changes in the dominant nutrient over space and time (nitrogen versus phosphorus during primary succession). In this section, we present several models that examine how incorporating multiple species or multiple nutrients can lead to novel effects on stability and functioning in ecosystems. For spatial networks, we can look at how moving towards finite topologies can lead to the realization that the connectivity and the structuring of movement play separate roles in regulating and influencing functioning and stability within meta-ecosystems. A portion of these properties can be elucidated from the spectra of the connectedness matrix, which can be conceptualized as the scales of spatial interactions. Finally, we cover ways of integrating the effects of numerous materials within ecosystems that do not participate in trophic interactions directly, and yet that can have large impacts on ecosystem functioning and stability. From examples such as materials used as structure by foundation species, we construct a general model of non-trophic material fluxes in meta-ecosystems.

4.1 Introduction

In the previous chapters, we developed relatively simple models of meta-ecosystems. The models were either spatially implicit [58, 109] or had two-patches [54, 57, 101, 115]. The models also focused on a single nutrient that moved among ecosystem compartments and we generally restricted the ecosystem to have a single compartment

© The Author(s), under exclusive license to Springer Nature Switzerland AG 2021   55
F. Guichard and J. Marleau, *Meta-Ecosystem Dynamics*, Lecture Notes
on Mathematical Modelling in the Life Sciences,
https://doi.org/10.1007/978-3-030-83454-8_4

per trophic level. While these simplifications can aid in the analysis of models and provide important insights, many important ecosystem functions cannot be captured by such limiting assumptions.

In this chapter, we will be focusing our efforts in order to generalize across spatial topologies, species diversity, multiple nutrients and non-resource materials. For exploring topologies, we will formally introduce the concepts of physical connectedness and spatial flux rate matrices, which allow us to describe the connectivity between ecosystems. Furthermore, we will show that the properties of the physical connectedness matrix can generate novel spatiotemporal dynamics that cannot occur in two-patch meta-ecosystem models [116].

As heterogeneity in spatial fluxes between compartments can alter ecosystem properties, it leads naturally to consider heterogeneity within a compartment. Species diversity within a given trophic level brings us closer to community ecology and provides an opportunity to explore biodiversity-ecosystem functioning patterns and processes in an explicitly ecosystem context. We illustrate how small differences in spatial fluxes between competing primary producers along with differences in assimilation efficiencies can lead to regional-scale differences in inorganic nutrients, promoting coexistence and altering ecosystem functioning [114].

Since multiple nutrients requires explicit consideration of the balance or ratio of the nutrients to determine which elemental nutrient may limit growth, the stoichiometries of the ecosystem compartments will need to be explicitly considered [117]. We begin by presenting a relatively simple ecosystem model inspired by [33, 34], which serves to introduce the concept of nutrient limitation. We then show multiple nutrients can possibly limit growth at the same time, which is difficult to achieve in the [34] and other stoichiometric models, through spatial fluxes of nutrients and organisms [117]. We also consider how these fluxes can also impact competitive dynamics in meta-ecosystems [167].

Finally, we look beyond resource materials as the only kind of material fluxes in meta-ecosystems. While fluxes of nitrogen, phosphorus and other limiting nutrients are fundamental in ecosystem studies, fluxes of substances such as mercury can have massive impacts on the functioning and structure of ecosystems. Furthermore, organisms can generate non-resource materials that can alter the physical and chemical environment of their habitats through ecosystem engineering [30, 31, 88], which can drastically alter resource fluxes [87, 96]. Adding these elements to meta-ecosystem models represents a fundamental challenge, as the impact of substances can change drastically between environments and organisms [118]. We suggest ways of extending meta-ecosystem models to capture these impacts, though this is a rapidly developing field of investigation.

4.2 From Two to Many: Multi-patch Spatial Topologies in Meta-Ecosystems

In our earlier chapters, when connecting two ecosystems together through spatial fluxes, we utilized the following coupling function $\mu_X(X_i, X_j) = -\mu_X(X_j, X_i) = m_X(X_j - X_i)$, where X is a generic ecosystem compartment, m_X is the spatial flux rate for that ecosystem compartment and i and j indices represent the two different ecosystems. However, what happens if we have three or more ecosystems in our meta-ecosystem? What should our coupling functions look like?

Before continuing, we should first consider two important properties of spatially-explicit ecological systems that we have so far kept together: movement and physical connectedness. We can illustrate the distinction by thinking of lakes coupled together by streams with weak directional flows (Fig. 4.1). Our ecosystems include the nutrients available in the water (available nutrients N), primary producers in the form of algae (B_P) and the herbivores that consume the the algae (primary consumers B_C). As a first approximation, herbivores are the only compartment that is connected between neighbouring patches in the landscape, thus $m_C > 0$. However, a herbivore in lake 1 cannot reach lake 4 directly, but must go through lakes 2 and 3 first (Fig. 4.1).

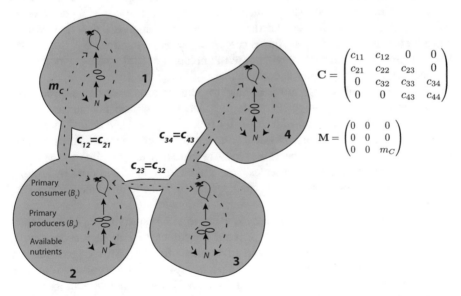

$$\mathbf{C} = \begin{pmatrix} c_{11} & c_{12} & 0 & 0 \\ c_{21} & c_{22} & c_{23} & 0 \\ 0 & c_{32} & c_{33} & c_{34} \\ 0 & 0 & c_{43} & c_{44} \end{pmatrix}$$

$$\mathbf{M} = \begin{pmatrix} 0 & 0 & 0 \\ 0 & 0 & 0 \\ 0 & 0 & m_C \end{pmatrix}$$

Fig. 4.1 Conceptual example of a four-lake aquatic nutrient (N)-primary producer (B_P)-primary consumer (B_C) meta-ecosystem. There are physical connections between lakes 1 and 2 (c_{12}), 2 and 3 (c_{23}) and 3 and 4 (c_{34}) and the connections are symmetrical (i.e. $c_{21} = c_{12}$). However, only one ecosystem compartment is able to move between ecosystems, the primary consumers, and thus have the only spatial flux rate m_C. The movement of primary consumers between ecosystems is then a product of local biomasses/concentrations, the matrix of spatial flux rates, $\mathbf{M}$, and the physical connectedness matrix $\mathbf{C}$

Therefore, there is no connection between ecosystem 1 and 4, thus $\mu_X(X_1, X_4) = 0$ for all ecosystem compartments.

We can formalize this distinction between spatial flux rates and the physical connectedness of the meta-ecosystem by introducing two matrices, the spatial flux rate matrix $\mathbf{M}$ and the connectedness matrix $\mathbf{C}$. The spatial flux rate matrix is a square k by k matrix, where k is the number of different types of ecosystem compartments found in the meta-ecosystem. The diagonal entries of the matrix are the spatial flux rates of each compartment, while the off-diagonal entries represent cross-compartment spatial flux rates where the biomass and/or concentration of one compartment influences the spatial fluxes of another. For example, if a consumer commonly transports large amounts of detritus with it as it moves between ecosystems, then we would utilise a cross-compartment spatial flux rate to capture the movement of detritus caused by the consumers.

For the connectedness matrix, which in earlier work is also known as the connectedness matrix, we have a n by n matrix, where n is the number of ecosystems in the meta-ecosystem. In most meta-ecosystem models [54, 115–117, e.g.], there is an assumption of symmetry, i.e. the connections are bidirectional and equal such that $c_{ij} = c_{ji}$ and therefore $\mathbf{C} = \mathbf{C}^T$, ^{T} is the transpose. Therefore, the interpretation of the off-diagonal entries is simply the presence of a connection. It is also commonly assumed that the spatial flux out of the ecosystem is equal to the spatial fluxes received by the other ecosystems, such that $m_X \sum_{j=1}^{n} c_{ji} = m_X \sum_{j=1}^{n} c_{ij} = 0$. At the end of this chapter, we will relax these assumptions, which will then require us to consider the direction of spatial fluxes (i.e. $m_X c_{ij} \neq m_X c_{ji}$ and $m_X \sum_{j=1}^{n} c_{ji} \neq 0$). Under these circumstances, we will adopt the notation of [86] where c_{ji} indicates a connection from ecosystem j to ecosystem i.

With this formalism, we are now able to address a number of questions that are of particular interest to spatial ecologists. Does the physical connectedness of the meta-ecosystem have an impact on its functioning and dynamics? Are there important interactions between physical connectedness and the rates of spatial fluxes that determine the kinds of impacts we see? Are there metrics that allow us to predict these impacts at the right spatial scales? We will address these questions by looking at how the addition of physical connectedness can impact a familiar meta-ecosystem model.

If we recall the [115] model in Chap. 2, we have a inorganic nutrient compartment N_i, a primary producer $B_{P,i}$ and a primary consumer $B_{C,i}$ in each ecosystem i, though we used nutrient stock instead of biomass for the primary producer and consumer ($N_{P,i}$ and $N_{C,i}$, respectively). Two ecosystems were connected through the spatial flux of inorganic nutrients, the flux was diffusive and without loss, such that $\mu_N(N_i, N_j) = -\mu_N(N_j, N_i) = m_N(N_j - N_i)$.

This model was generalized in [116] by allowing movement for the other compartments and by having more than two ecosystems, which gives rise to the following system of equations:

$$\frac{dN_i}{dt} = I_N - E_N N_i - \frac{\alpha_P N_i N_{P,i}}{\beta_P + N_i} + \epsilon_P l_P N_{P,i} + \epsilon_C l_C N_{C,i} + \gamma \frac{\alpha_C N_{P,i} N_{C,i}}{\beta_H + N_{P,i}}$$

$$+ m_N \sum_{j=1}^{n} c_{ij} N_j \tag{4.1a}$$

$$\frac{dN_{P,i}}{dt} = \frac{\alpha_P N_i N_{P,i}}{\beta_P + N_i} - l_P N_{P,i} - \frac{\alpha_C N_{P,i} N_{C,i}}{\beta_C + N_{P,i}} + m_P \sum_{j=1}^{n} c_{ij} N_{P,j} \tag{4.1b}$$

$$\frac{dN_{C,i}}{dt} = (1 - \gamma) \frac{\alpha_C N_{P,i} N_{C,i}}{\beta_C + N_{P,i}} - l_C N_{C,i} + m_C \sum_{j=1}^{n} c_{ij} N_{C,j} \tag{4.1c}$$

where the parameters have the same meaning as in Chap. 2, except we also have a γ parameter that represents the amount of nutrients that are egested by the primary consumer and this impacts quantitative, though not qualitative, results of the model [116]. As in Chap. 2, we will focus primarily on the stability of the spatially homogeneous solution of interest, which in our case is an equilibrium (E_c) where we have positive amounts of nutrients, primary producers and herbivores.

When there is no spatial flux between ecosystems, the local stability of E_c in a given ecosystem i can be derived from its associated Jacobian matrix, $\mathbf{J}$. To simplify notation, we let $dN_i/dt = F(N_i, N_{P,i}, N_{H,i})$, $dN_{P,i}/dt = G(N_i, N_{P,i}, N_{H,i})$ and $dN_{C,i}/dt = K(N_i, N_{P,i}, N_{H,i})$, which gives us:

$$\mathbf{J} = \begin{pmatrix} \frac{\partial F}{\partial N_i} & \frac{\partial F}{\partial N_{P,i}} & \frac{\partial F}{\partial N_{C,i}} \\ \frac{\partial G}{\partial N_i} & \frac{\partial G}{\partial N_{P,i}} & \frac{\partial G}{\partial N_{C,i}} \\ \frac{\partial K}{\partial N_i} & \frac{\partial K}{\partial N_{P,i}} & \frac{\partial K}{\partial N_{C,i}} \end{pmatrix} = \begin{pmatrix} j_{11} & j_{12} & j_{13} \\ j_{21} & j_{22} & j_{23} \\ j_{31} & j_{32} & j_{33} \end{pmatrix} \tag{4.2}$$

where the j_{ik} parameters represent the effect of an increase in ecosystem compartment k at equilibrium on ecosystem compartment i. For example, j_{21} is the effect of increasing available nutrient N on primary producer nutrient N stock at equilibrium. Generally, it is assumed that any increase in limiting nutrient should increase primary producer biomass (and hence nutrient stock), which means that for most ecosystem models j_{21} has a positive value. Similar derivations of the signs for the j_{ik} parameters in the Jacobian can be done, but will not be discussed in detail here. E_c is locally stable if the following three conditions are met:

$$-\mathbf{tr}(\mathbf{J}) > 0 \tag{4.3a}$$

$$-\mathbf{det}(\mathbf{J}) > 0 \tag{4.3b}$$

$$-\mathbf{tr}(\mathbf{J}) * (\mathbf{J}_{11} + \mathbf{J}_{22} + \mathbf{J}_{33}) > -\mathbf{det}(\mathbf{J}) \tag{4.3c}$$

where $\mathbf{tr}$ and $\mathbf{det}$ are the trace and determinant of the matrix, respectively, while $\mathbf{J}_{11}$, $\mathbf{J}_{22}$ and $\mathbf{J}_{33}$ are the cofactors of the Jacobian matrix. For the first condition to be met, it requires at least one compartment to experience self-limitation in growth at equilibrium, and that self-limitation needs to be stronger than any positive feedbacks found in the other compartments. The second and third conditions are too complex to understand biologically without some knowledge of the signs of j_{ik} parameters.

Our specified equations result in j_{11}, j_{12} and j_{23} being negative while j_{13}, j_{21}, j_{22} and j_{32} being positive at equilibrium E_c. Furthermore, j_{31} and j_{33} will be equal to zero at equilibrium E_c. Using the information about the signs of the Jacobian matrix elements and the stability conditions in Eq. 4.3, we can derive the following relationships between the j_{ik} that ensure the equilibrium E_c will be locally stable:

$$|j_{11}| > j_{22} \tag{4.4a}$$

$$|j_{11}j_{23}| > |j_{13}j_{21}| \tag{4.4b}$$

$$j_{11}(-j_{11}j_{22} + j_{12}j_{21}) + j_{13}j_{21}j_{32} > |j_{22}(j_{23}j_{32} - j_{11}j_{22} + j_{12}j_{21})| \tag{4.4c}$$

Our interest lies in how the addition of movement and spatial structure to the local ecosystem model can destabilize the equilibrium E_c without any changes to the j_{ik} parameters. Technically, we could try to examine the eigenvalues of the Jacobian matrix for the full meta-ecosystem model in Eq. 4.1, but this matrix would be a very large (nk x nk) and difficult to analyze. However, [86] proved that it is possible to instead analyze n decoupled k x k matrices (i.e. the size of $\mathbf{J}$) instead. Each matrix $\mathbf{V}(i)$ ends up being equal to:

$$\mathbf{V}(i) = \mathbf{J} + \lambda_i \mathbf{M}$$

where λ_i is the ith eigenvalue of the connectedness matrix. Because the properties of the connectedness matrices used here, the largest λ is equal to zero and corresponds to the no spatial flux case, and the rest of the eigenvalues are constrained between 0 and -2. These eigenvalues help capture the spatial scale at which the meta-ecosystem responds to perturbations, which is why they can also be called the 'scales of spatial interaction' [116]. To be more precise, the spatial scale is the network of ecosystems, such that certain patterns will emerge in one, two or more ecosystems and these patterns are related to the eigenvalues.

For the spatially homogeneous solution to hold across the metaecosystem, i.e. all ecosystems are at equilibrium E_c, all the eigenvalues of each matrix $\mathbf{V}(i)$ must have negative real parts. For this to be the case, the conditions described in Eq. 4.3 must hold for each $\mathbf{V}(i)$. We therefore have:

$$-\mathbf{tr}(\mathbf{V}(i)) = -\mathbf{tr}(\mathbf{J}) - \lambda_i(m_N + m_P + m_C) > 0 \tag{4.5a}$$

$$-\mathbf{det}(\mathbf{V}(i)) = -\mathbf{det}(\mathbf{J}) + j_{23}j_{32}\lambda_i m_N - \lambda_i m_C(v_{11}v_{22} - j_{12}j_{21}) > 0 \tag{4.5b}$$

$$\mathbf{det}(\mathbf{V}(i)) > \mathbf{tr}(\mathbf{V}(i)) * (\mathbf{V}_{11} + \mathbf{V}_{22} + \mathbf{V}_{33}) \tag{4.5c}$$

where $v_{11} = j_{11} + \lambda_i m_N$ and $v_{22} = j_{22} + \lambda_i m_P$. Note that the inequality in (4.5a) will always be satisfied as λ_i must be equal to or less than zero. Therefore, only inequalities (4.5b) and (4.5c) can be violated. From inequality (4.5b), we can quickly derive a value of m_C at which the inequality will no longer hold by setting the left-hand size of the inequality to zero:

$$-\mathbf{det}(\mathbf{J}) + j_{23}j_{32}\lambda_i m_N - \lambda_i m_C(v_{11}v_{22} - j_{12}j_{21}) = 0$$
$$\Leftrightarrow \quad \lambda_i m_C(v_{11}v_{22} - j_{12}j_{21}) = -\mathbf{det}(\mathbf{J}) + j_{23}j_{32}\lambda_i m_N$$
$$\Leftrightarrow \quad m_C = \frac{-\mathbf{det}(\mathbf{J})+j_{23}j_{32}\lambda_i m_N}{\lambda_i(v_{11}v_{22}-j_{12}j_{21})}$$

From the above derivation, we can define the critical consumer spatial flux rate function, which is a function of λ_i:

$$m_C^c(\lambda_i) = \frac{-\mathbf{det}(\mathbf{J}) + j_{23}j_{32}\lambda_i m_N}{\lambda_i(v_{11}v_{22} - j_{12}j_{21})} \tag{4.6}$$

Finally, we can define the minimal critical consumer spatial flux rate value, $m_C^{min,c}$ as being the minimum value of the critical consumer spatial flux rate function: $m_C^{min,c} = \min(m_C^c(\lambda_2), m_C^c(\lambda_3), ..., m_C^c(\lambda_n))$. Thus, whenever we have a value of m_C greater than $m_C^{min,c}$, E_c will be destabilized by the spatial flux of consumers. What happens after this destabilization, which is caused by a fold bifurcation (i.e. one eigenvalue associated with the equilibrium transitions from having negative real part to a positive real part), is the creation of spatially heterogeneous equilibria across the meta-ecosystem (i.e. the equilibrium values differ between each local ecosystem; see Sect. 4.3 for an example).

From inequality (4.5c), it is possible to derive a value of m_N at which the inequality no longer holds, though we will just present the final product here:

$$m_N = \frac{-\zeta-\sqrt{\zeta^2+4\lambda_i^2(v_{22}+v_{33})\xi}}{-2\lambda_i^2(v_{22}+v_{33})}$$

$$\zeta = \zeta(\lambda_i, m_P, m_C) = -2\lambda_i j_{11}(v_{22} + v_{33}) + \lambda_i(j_{12}j_{21} - v_{22}^2 - 2v_{22}v_{33} - v_{33}^2)$$
$$\xi = \xi(\lambda_i, m_P, m_C) = -j_{11}^2(v_{22} + v_{33}) + j_{11}(j_{12}j_{21} - v_{22}^2 - 2v_{22}v_{33} - v_{33}^2) + $$
$$v_{22}(j_{12}j_{21} + j_{23}j_{32} - v_{22}v_{33}) + v_{33}(j_{23}j_{32} - v_{22}v_{33}) + j_{13}j_{21}j_{32}$$

where $v_{33} = \lambda_i m_C$. We can define the critical nutrient spatial flux rate function, which is a function of λ_i:

$$m_N^c(\lambda_i) = \frac{-\zeta - \sqrt{\zeta^2 + 4\lambda_i^2(j_{22} + \lambda_i(m_P + m_C))\xi}}{-2\lambda_i^2(j_{22} + \lambda_i(m_P + m_C))} \tag{4.7}$$

We can now define the minimal critical nutrient spatial flux rate value, $m_N^{min,c}$ as being the minimum value of the critical nutrient spatial flux rate function: $m_N^{min,c} = \min(m_N^c(\lambda_2), m_N^c(\lambda_3), ..., m_N^c(\lambda_n))$. Thus, whenever we have a value of m_N greater than $m_N^{min,c}$, E_c will be destabilized by the spatial flux of nutrients. What happens after this destabilization, which is caused by a spatial Hopf bifurcation (i.e. when two conjugate eigenvalues with imaginary parts of a stable equilibrium transition from having negative real parts to positive real parts), is the creation of

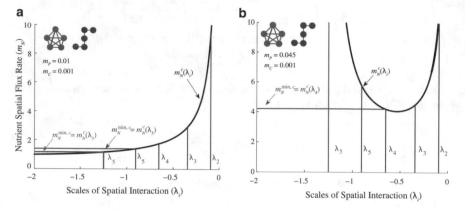

Fig. 4.2 The effects of spatial configuration and spatial flux rates on the critical nutrient spatial flux rate at **a** low and **b** high consumer spatial flux rates. The black line represents the critical nutrient spatial flux rate function evaluated at all possible connectedness matrix eigenvalues, while the vertical red and blue lines are the discrete eigenvalues of the inset 5-ecosystem meta-ecosystems. Adapted from [116]

spatially heterogeneous limit cycles across the meta-ecosystem (see Sect. 3.4.1.1 for an example).

The importance of multiple λ_i values occurs due to the non-linear, non-monotonic relationship between λ_i and the critical spatial flux functions (Fig. 4.2). For example, at low spatial flux rates of the primary producer and consumer, the $m_N^c(\lambda_i)$ decreases with decreasing λ_i, thus $m_N^{min,c}(\lambda_i) = m_N^c(\lambda_5)$ for 5-ecosystem meta-ecosystems (Fig. 4.2a). Under this scenario, heavily connected meta-ecosystems will generally have smaller λ_i values and thus be easier to destabilize through spatial fluxes (Fig. 4.2a).

Furthermore, which scale of spatial interaction is associated to the destabilization will impact local and meta-ecosystem level dynamics and functions. Since each λ_i is associated with a unique eigenvector, the dynamics after destabilization closely mimic the synchronization patterns of that eigenvector for values near the bifurcation, leading to unique regional dynamics (Fig. 3.12). These differences in dynamics also lead to differences in ecosystem function between different meta-ecosystems configurations after bifurcation, despite having identical local parameters (Fig. 4.3). Thus, seemingly small changes in meta-ecosystem connectivity can lead to large-scale alterations in meta-ecosystem dynamics and functioning, proving the importance of considering spatial configuration explicitly in meta-ecosystem theory.

Fig. 4.3 The effects of spatial configuration (inset images and associated lines) and spatial flux rates on regional consumer production. Adapted from [116]

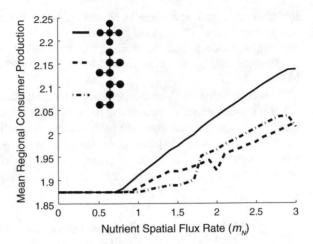

4.3 Beyond the Chain: Species Diversity in Meta-Ecosystems

As we have seen, a diversity of spatial configurations along with heterogeneity in spatial flux rates between different ecosystem compartments can drastically alter meta-ecosystem dynamics and functioning. However, the meta-ecosystem models explored so far have been highly restrictive in one sense: there is no local diversity within a given ecosystem compartment. This homogenization, while convenient analytically, prevents us from both examining the effects of diversity on ecosystem dynamics and function as well as examining how spatial ecosystem dynamics and function promote diversity. In this section, we will focus on how diversity within one of the biotic compartments can lead help us understand how spatial ecosystem functions promote coexistence.

The impacts of species diversity on ecosystems is the primary focus of what is known as 'biodiversity and ecosystem functioning' research [166]. Since the late 1990s, ecologists have performed numerous experiments examining how increasing species richness can lead to increases in primary production, biomass and other ecosystem properties [166]. However, it is not clear what processes are driving these changes as mechanisms of coexistence are rarely studied directly [44]. Statistical methods that have been used to link ecosystem properties to sets of mechanisms [48, 108, e.g.] have been criticized due to their inability to distinguish non-linear population level effects from community-level mechanisms [144]. One way of addressing these issues is to look at how ecosystem and community processes can drive coexistence, and look at how the mechanism of coexistence impacts the ecosystem-level function.

For example, a consumer can allow two resource species to coexist on a single limiting nutrient by preferentially consuming the better nutrient competitor, resulting in a trade-off between nutrient use and consumption avoidance [59]. However, this trade-off is extremely fragile, as small changes in nutrient supply or the ecosystem nutrient turnover rate (i.e. how fast do nutrients move out of the ecosystem) prevent coexistence from occurring [59, Fig. 4.4a]. Thus, this consumer-mediated coexistence mechanism was viewed as being unlikely to greatly contribute to biodiversity [59].

This conclusion was based on looking at local, spatially homogeneous processes. What if the consumer, resources and the limiting nutrient were in a spatial environment? To investigate the potential for herbivores to facilitate autotroph coexistence in meta-ecosystems, [114] used the parameters and functions specified by [59] for their phosphorus (N)-two algae (B_{P_w}-*Daphnia* (C) model to create a meta-ecosystem model defined by the following system of differential equations:

Fig. 4.4 The parameter ranges of nutrient supply and dilution rate that allow for consumer-mediated coexistence (red areas) between two primary producers limited by the same nutrient in the **a** non-spatial and **b** spatial ecosystem models. Adapted from [114]

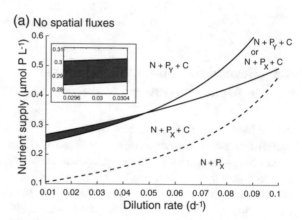

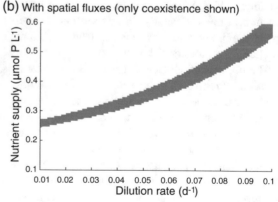

$$\frac{dN_i}{dt} = \mathcal{D}S - \mathcal{D}N_i - \sum_{w=1}^{k} q_w \left(\frac{r_w N_i B_{P_{wi}}}{K_w + N_i} - l_w B_{P_{wi}} \right) + q_C bl_C C_i$$

$$+ \sum_{w=1}^{k} \frac{(q_w - b\epsilon_w q_C) a_{Cw} B_{P_{wi}} C_i}{1 + a_{C1} h_{C1} B_{P_{1i}} + \dots + a_{Ck} h_{Ck} B_{P_{ki}}} + \sum_{j=1}^{n} c_{ij} m_N N_j \qquad (4.8a)$$

$$\frac{dB_{P_{wi}}}{dt} = \frac{r_w N_i B_{P_{wi}}}{K_w + N_i} - (l_w + \mathcal{D}) B_{P_{wi}} - \frac{a_{Cw} B_{P_{wi}} C_i}{1 + a_{C1} h_{C1} B_{P_{1i}} + \dots + a_{Ck} h_{Ck} B_{P_{ki}}}$$

$$+ \sum_{j=1}^{n} c_{ij} m_{P_w} B_{P_{wj}} \qquad (4.8b)$$

$$\frac{dC_i}{dt} = \sum_{w=1}^{k} \frac{\epsilon_w a_{Cw} B_{P_{wi}} C_i}{1 + a_{C1} h_{C1} B_{P_{1i}} + \dots + a_{Ck} h_{Ck} B_{P_{ki}}} - (l_C + \mathcal{D}) C_i$$

$$+ \sum_{j=1}^{n} c_{ij} d_C C_j \qquad (4.8c)$$

While most of the variables and parameters are already familiar, we need to high-light some key differences to earlier models. Each local ecosystem i behaves as an chemostat, which introduces a constant dilution rate (i.e. turnover rate of the volume of the chemostat), $\mathcal{D}$, that applies to all ecosystem compartments. Furthermore, the supply of phosphorus, S, is also constant and becomes the supply rate when multiplied by the dilution rate. In this model, the number of consumers, C_i, is tracked while biomass is derived by multiplying C_i by b, the molar carbon in one individual consumer. For the k possible primary producers, each grows according to a Monod growth function with r_w being the maximal growth rate and K_w being the half-saturation constant. a_{Cw} and h_{Cw} are the attack rate and handling time, respectively, of the consumer feeding on a primary produce w, which allows us to model the impacts of multiple primary producers on the modified type II functional response. Finally, note that all nutrients lost due to imperfect assimilation or mass-specific loss outside of the dilution rate are recycled, which imposes a mass balance constraint on the system ().

By allowing spatial fluxes between multiple 'chemostat' ecosystems leads to a massive expansion of the parameter range allowing for coexistence [114, Fig. 4.4b]. This increase in coexistence is due to the creation of very different local ecosystems, as the spatial fluxes allow for the creation of 'high nutrient supply' and 'low nutrient supply' ecosystems, where nutrient supply includes both the supply rate and the nutrients recycled by the consumer. The consumer-resistant primary producer dominates (P_y) the high nutrient supply ecosystem and the superior nutrient competitor primary producer (P_x) dominates the low nutrient supply ecosystem [114]. This gradient in nutrient supply is developed differences in assimilation efficiency between primary producers, such that more nutrients are recycled back after the consumer eats P_y, which then promotes its growth at the expense of P_x, while if P_x is consumed, less nutrients are recycled, which helps it maintain is advantage over P_y (Fig. 4.5).

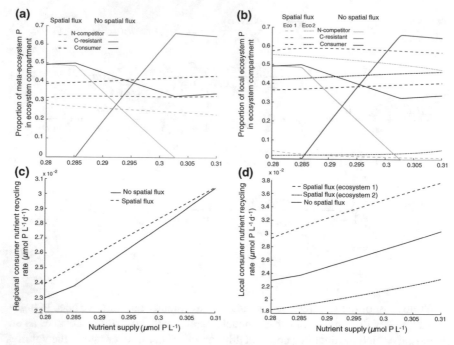

Fig. 4.5 Changes in meta-ecosystem and local ecosystem structure and function based on coexistence mechanisms across a gradient of nutrient supply. Parameter values are $D = 0.03$, $m_N = m_{P_x} = m_{P_y} = 0.001$ with $m_C = 0.5$ with movement and $m_C = m_C = m_{P_x} = m_{P_y} = 0$ without movement for all panels. **a** The relative distribution of phosphorus in the meta-ecosystem among consumers (black lines), the superior nutrient competitor (green lines) and consumer-resistant (blue lines) algae with (dashed lines) and without (solid lines) movement. **b** The relative distribution of phosphorus within local ecosystems with (dashed and dash-dotted lines) and without (solid lines) movement. **c** Rate of consumer nutrient recycling at the level of the meta-ecosystem with (dashed line) and without (solid line) movement. **d** Rate of consumer nutrient recycling at the level of local ecosystems with (dashed and dash-dotted lines) and without (solid line) movement. Adapted from [114]

This mechanism of coexistence requires spatial fluxes between the ecosystems to allow for some of the ecosystems to be a sink for the primary producers, without there being too much flux that negatively impacts the source that it alters local ecosystem functioning [114]. Thus, if local ecosystem conditions are too rich in nutrients or if the rates of loss due to dilution are too large, the spatial gradients cannot be maintained at the dynamics and functioning revert to local ecosystem model predictions, which involves competitive exclusion. However, high dilution rates in general are beneficial in promoting coexistence, which is contrary to local ecosystem predictions (Fig. 4.6c).

Furthermore, because the coexistence equilibria is created by fold bifurcations, the equilibria that involve competitive exclusion are still locally stable (Fig. 4.6a, b, solid red and blue lines). Because there are multiple stable equilibria, the initial conditions will then determine the trajectory of the meta-ecosystem towards either coexistence or competitive dominance by one species (Fig. 4.6a, b). Thus, the coexistence created by the redistribution of organisms and nutrients across the meta-ecosystem is not necessarily robust to any perturbation.

These results from [114] illustrate the importance of spatial scale and of mechanisms in understanding how biodiversity interacts with ecosystem function [44, 51]. For our given function of nutrient recycling, we see that local coexistence caused by the spatial redistribution of nutrients and organisms could lead to either an increase or decrease in recycling compared to a lack of coexistence or if coexistence was mediated purely by local mechanisms (Fig. 4.5d). Nevertheless, coexistence due to spatial mechanisms always led to more recycling than the equivalent spatially homogeneous meta-ecosystem (Fig. 4.5c). Thus, knowing that there is greater diversity in ecosystem 1 versus ecosystem 2 is insufficient to predict higher or lower functioning without knowledge of its spatial processes.

4.4 More Than Carbon: Multiple Elemental Nutrients in Meta-Ecosystems

Another form of diversity, though not of a biological kind, is the elemental stoichiometry of ecosystems. Throughout this text, we have focused on only one 'currency' of interest at a time, such as energy or a limiting nutrient [118]. In many regards, this approach has been empirically validated for a number of ecosystems, including freshwater lakes [158]. However, there is also substantial experimental evidence that primary production can be limited by multiple nutrients [67]. Furthermore, it has long been known that the ratios of elements in organisms, in soil and in water can vary greatly and lead to stoichiometric imbalances that can control ecosystem processes [111, 149, 161]. Determining when and where multiple limiting nutrients need to be considered in understanding ecosystem processes requires us to consider stoichiometrically-explicit models.

One relatively simple way of building a stoichiometric ecosystem model is to think of the (dry) biomass of an organism being divided up into amounts of various elements [33, Fig. 4.7a]. For example, primary producer biomass B_P is made up of an amount of carbon, $\mathcal{P}_{carbon}$, of phosphorus, $\mathcal{P}_{phosphorus}$, of nitrogen, $\mathcal{P}_{nitrogen}$, etc. We could then say that $B_P = \sum_{k=1}^{l} \mathcal{P}_k$, when we have l different elements. Now, if these elements are kept in strict proportion (i.e. fixed stoichiometry), we can derive fixed constant quotients for each element k in the primary producer:

$$q_k = \frac{\mathcal{P}_k}{\sum\limits_{k=1}^{l} \mathcal{P}_k} \tag{4.9}$$

Similarly, if we have a consumer in the ecosystem, we can also define the amount for a given element k, Θ_k, and then derive fixed constant quotients:

$$\rho_k = \frac{\Theta_k}{\sum\limits_{k=1}^{l} \Theta_k} \tag{4.10}$$

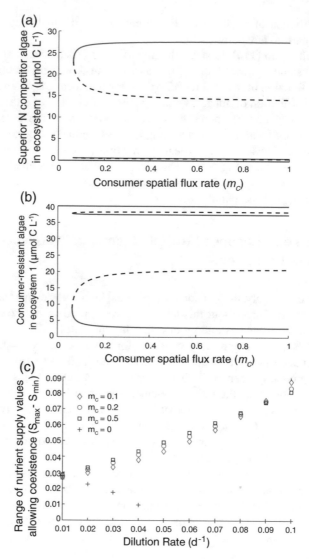

Fig. 4.6 Spatially heterogeneous coexistence of algae outside the parameter space that allows for spatially homogeneous coexistence. $\mathcal{D} = 0.03$ and $m_N = m_{P_x} = m_{P_y} = 0.001$ for all panels. **a** and **b** Bifurcation diagrams with $S = 0.31$ for the (a) consumer-resistant algae (P_y) and the (b) superior competitor algae (P_X) in ecosystem 1 with the consumer spatial flux rate, m_C, as the bifurcation parameter. The lines indicate the existence of an equilibrium, with the red line indicating a P_y-consumer spatially homogeneous equilibria, which is stable, the blue lines indicating the spatially heterogeneous coexistence equilibria, which are also stable, that are separated by unstable equilibria (dashed black lines). **c** Quantification of the range of S values ($S_{max} - S_{min}$) that allow for coexistence with no consumer movement (+), and at $m_C = 0.1$ (open diamond), $m_C = 0.2$ (open circle) and $m_C = 0.5$ (open square). Adapted from [114]

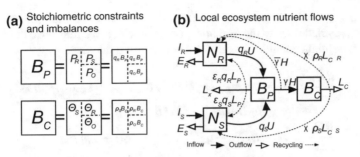

Fig. 4.7 The structure of ecosystem stoichiometric models. **a** Each biotic ecosystem compartment's biomass (e.g. B_P) can be subdivided into elemental stocks (e.g. P_R, P_S), and those that do not limit the growth of the organism can be lumped together as other elements (O). Note that the relative ratios of each element differ between the consumer (B_C) and primary produce (B_P in this example. **b** Within local ecosystems, the fluxes of nutrients between compartments are therefore dependent on the stoichiometries of the organisms, as they assimilate maximally the most limiting nutrient while excreting or limiting their uptake of the non-limiting nutrient. Symbols originate from Eq. 4.11. Adapted from [117]

For most stoichiometric models, there is usually is an assumption that only two elemental nutrients will be limiting [33, 34, 90, 91, 173, e.g.]. Therefore, we will limit our examination to two limiting nutrients, R and S, with all the others being part of the 'other' elements and assumed to be non-limiting. We will also make the arbitrary decisions that the amount of those non-limiting elements are equal in consumers and primary producers, and that the consumer has a lower $R:S$ ratio than the primary producers, which implies $q_R > \rho_R$ and $\rho_S > q_S$.

Moving from away from organismal stoichiometry to the local ecosystem, stoichiometric models have a number of features in common with single nutrient models, yet also a number of differences (Fig. 4.7b). As in earlier models, influx of nutrients into the inorganic stocks are at a constant rate (I_R and I_S), while efflux of available nutrients are proportional to current nutrient stock ($E_R N_R$ and $E_S N_S$). It should be noted that many models make use of the 'chemostat' approximation for ecosystems, which results in a constant amount of total nutrients in the ecosystem at equilibrium and is called the 'supply point' [34, e.g.]. To emphasize this point, it is common to parse the influx rates into two quantities, the dilution rate constant ($\mathcal{D}$, 1/time), and the supply point ($\mathcal{S}$, g or mol nutrient). If the efflux rate constants and all other losses are equal to the dilution rate constant, then the chemostat conditions hold for all nutrients.

Available nutrients are assimilated by the primary producers according to a nutrient uptake function $U(N_R, N_S, B_P)$, who are in turn ingested by the consumers, $H(B_P, B_Y)$, though only a portion, γ, is successfully assimilated and the rest return back to the available pool (Fig. 4.7b). Since we assumed that the $R : S$ ratio of the consumers was lower than that of primary producers, we only have excess nutrient R recycled in this way. Nutrients are also lost from the primary producers and consumers according to loss functions, $L_P(B_P)$ and $L_C(B_C)$, respectively, with only a proportion, ϵ_R and ϵ_S for primary producers and χ_R and chi_S for the consumer, of the nutrients recycled back to the available pool (Fig. 4.7b). We thus have the following set of differential equations for our local ecosystem dynamics

$$\frac{dN_R}{dt} = I_R - E_R N_R - q_R U(N_R, N_S, B_P) + \epsilon_R q_R L_P(B_P)$$

$$+ \chi_R \rho_R L_C(B_C) + \bar{\gamma} H(B_P, B_C) \tag{4.11a}$$

$$\frac{dN_S}{dt} = I_S - E_S N_S - q_S U(N_R, N_S, B_P) + \epsilon_S q_S L_P(B_P)$$

$$+ \chi_S \rho_S L_C(B_C) \tag{4.11b}$$

$$\frac{dB_P}{dt} = U(N_R, N_S, B_P) - L_P(B_P) - H(B_P, B_C) \tag{4.11c}$$

$$\frac{dB_C}{dt} = \gamma H(B_P, B_C) - L_C B_C \tag{4.11d}$$

The specific forms used for these equations vary from model to model, though there are some commonalities. U is generally of the form $U(N_R, N_S)B_P$ as any non-linear effects of primary producer biomass on uptake is ignore. In order to operationalize Liebig's law of the minimum, i.e. the resource that is relatively the least available controls the growth of the organism, many models utilize a minimum function such that $U(N_R, N_S) = min(u_R(N_R), u_S(N_S))$. Another commonality is that loss rates are nearly always density-independent, such that $L_C(B_C) = l_C B_C$ and $L_P(B_P) = l_P B_P$, where l_C and l_P are rate constants.

In [34], they used Lotka-Volterra nutrient uptake functions and donor-controlled consumption function to examine how herbivores can shift the limitation status of the primary producers. Under equilibrium conditions, they showed that if the primary producers had greater affinity for R than S (defined as the ratio of uptake rate constant for R and S), the consumer could only promote a shift towards S limitation [34]. However, if the primary producer had greater affinity for S, then consumers could promote either S or R limitation, depending on the supply points and the exact $R : S$ of the consumer [34]. Consumers have this effect by altering the availability of both nutrients through biased nutrient recycling (i.e. they keep more nutrient R than nutrient S in their tissues) and through their consumption which moves some nutrients into consumer biomass, alters primary producer uptake and increases the minimal nutrient values necessary to maintain viable primary producer populations.

It must be noted that in the [34] model, what are called 'ultimate' limitation (i.e. what nutrient controls long-term production of biomass) and 'proximate' limitation (i.e. nutrient that controls short-term productivity of organism) are aligned [172]. This phenomenon occurs due to the donor-control function, which allows the primary producer to permanently gain biomass from a nutrient increase. With recipient-control (e.g. Lotka-Volterra or type II functional responses), there is no ultimate limitation of primary producers by nutrients as their biomass does not change with increasing nutrients at equilibrium. Thus, we only can only have proximate limitation.

However, models like this are not capable of reproducing many of the patterns of response of primary producers in semi-natural and experimental settings to multiple nutrient additions [67]. In many nitrogen and phosphorus addition experiments, researchers have observed that primary producers only respond to the addition of both nutrients (i.e. simultaneous colimitation) or they can respond to either nutrient (i.e. independent colimitation). Furthermore, the metric used to quantify this responses is

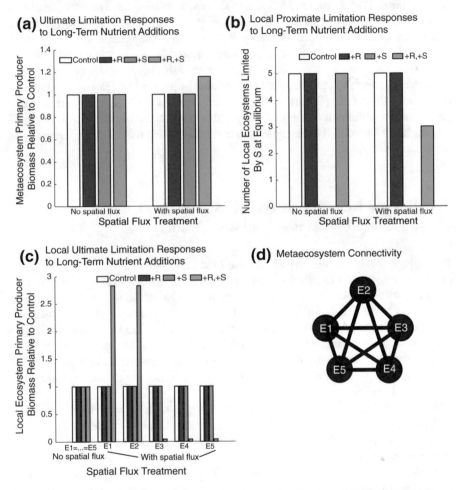

Fig. 4.8 How nutrient colimitation can emerge at local and regional spatial scales and at different time scales through spatial fluxes of nutrients and consumers. **a** Response to permanent nutrient additions (R, S and $R+S$) without and with spatial fluxes, leading to ultimate limitation responses at the meta-ecosystem level. **b** Response to temporary nutrient additions without and with spatial fluxes, leading to proximate nutrient limitation resposnes at the meta-ecosystem level. **c**, **a** Response to permanent nutrient additions (R, S and $R+S$) without and with spatial fluxes, leading to ultimate limitation responses at the local ecosystem level. **d** The configuration of the meta-ecosystem. Adapted from [117]

typically biomass, which under the expected recipient-control of primary producers should not result in any increase in biomass when nutrients are added [60].

Surprisingly, meta-ecosystem theory can provide the solution these conundrums. With the right mix of spatial fluxes from nutrients and consumers, spatial heterogeneity can be generated that leads to both simultaneous (Fig. 4.8a) and independent colimitation (Fig. 4.8b) at the meta-ecosystem scale, though one is ultimate limitation and the other is proximate [117]. The local ecosystems can also demonstrate colimited responses (Fig. 4.8c), though there is also the possibility of a negative response to nutrients, which is also seen in the literature [67]. Thus, both of the oddities found in practice could be explained in theory, but not with non-spatial theory.

There is a great deal of work left to be done in extending ecological stoichiometry into meta-ecosystem theory. Some work has been started examining the role of spatial heterogeneity and spatial fluxes of nutrients in promoting primary producer in coexistence in practice [63] and in theory [167], but this work generally ignores key processes like nutrient recycling. There are also landscape approaches to stoichiometry that may be fruitfully integrated into meta-ecosystem theory as we increase the realism of the models [102].

4.5 Beyond Resource Effects of Matter: Towards a Generalized Meta-Ecosystem Model

Throughout this book, we have emphasized the importance of the abiotic compartments of ecosystems. The fluxes of matter, be they through recycling or movement across ecosystem boundaries, can lead to profound changes to the communities that depend on them. However, we have only focused on a small subset of the substances that impact the structure and functioning of ecosystems: resources. Organisms live in environments that are populated by all kinds of substances, organic and/or inorganic, biotically-derived and/or abiotically produced, that they do not directly use for growth, maintenance and reproduction, but can drastically alter their existence [118].

In this section, we present an in depth example of how matter can acts as a material, i.e. a substance that alters the physical and chemical environment, and we propose how matter as information, i.e. a substance that can have multiple states that an organism can recognize and can impact a process, can be integrated into meta-ecosystem theory. We conclude this chapter by presenting, in a highly abstract way, a possible way to construct a generalized meta-ecosystem model that combines all the aspects elucidated in the previous sections of this chapter.

4.5.1 *Matter as Material: Foundation Species and the Construction of Meta-Ecosystems*

One example of non-resource effect of matter in meta-ecosystems is provided by foundation species or ecosystem engineers [30, 31, 88]. An ecosystem engineer is an organism whose presence or activity alters its physical surroundings or changes the flow of resources, thereby creating or modifying habitats and influencing all associated species [88]. Ecosystem engineers that have a disproportionate influence on other species compared with their abundance are called foundation species and can constitute defining components in many ecosystems such as forests, marine kelp forests, salt marshes and mussel bed ecosystems [4]. There is a rich theory of ecological dynamics integrating the net effect of foundation species on species growth and abundance [31, 88]. Niche construction theories have extended the process of ecosystem engineering to evolutionary dynamics [95, 135]. However, few studies have integrated the explicit alteration of flows of matter by foundation species that results in the physical modification of their habitat. These flows of matter include both resources that are consumed and material that are instead used to change physical properties of habitats.

Coastal marine ecosystems have provided many examples of foundation species such as reef forming corals that can shape coastal geomorphology. Others such as salt marshes, seagrass meadows, kelp forests and mussel beds affect the hydrodynamics and promote sedimentation and retention of materials suspended in the water column [4]. Recent meta-ecosystem theory has focused on the non-resource role of matter on the dynamics of mussel bed ecosystems [87, 96].

Reference [87] model local mussel population dynamics using a mean-field formulation of disturbance-recovery dynamics [61, 176]. We discussed this study in Sect. 3.4.1.1 to illustrate phase-amplitude synchrony in meta-ecosystems. We will now describe the model into more details as an example of non-resource effect of spatial flows of matter on meta-ecosystem dynamics. Disturbance dynamics in mussel beds can be captured as the proportion of three discrete states, similarly to models of disease dynamics (Fig. 4.9a): *mussel-covered* (M), *unoccupied* (U), and *disturbed* (unoccupied and recently disturbed, U^D). Disturbed areas are unoccupied areas that are recent gaps in the mussel bed, characterized by damaged byssal threads of surrounding mussels that become vulnerable to further wave disturbance defined by a disturbance propagation function g that can be function of the amount of detritus. When mussels along disturbed areas stabilize at rate γ by securing their byssal thread attachment, the area becomes simply unoccupied and available for the recruitment of adults and juveniles based on a recruitment function $p(M)$ and colonization rate α, leading to mussel-covered area.

Ecosystem maturation is then described by the amount of (in)organic matter (E) sequestered by the mussel bed (feces, pseudofeces, detritus, and sediments). We assume the supply of (in)organic matter is constant (ρ) and is limited to mussel-covered area M. The rate of matter resuspension is proportional to mussel dislodgement by waves (disturbance) and to a constant leaking rate ω corresponding to the

Fig. 4.9 Ecosystem and population interactions implemented in the mussel bed model. **a** Ecosystem (local) model structure. **b** Meta-ecosystem model structure under larval dispersal (biotic connectivity, d_M) or transport of (in)organic matter (abiotic connectivity, d_E). For clarity, we only show biotic and abiotic spatial coupling from patch i to j, but there is analogous coupling from patch j to i in the meta-ecosystem model. For **a** and **b** solid, dashed and dotted lines denote explicit matter fluxes, effects on biological or ecological rates (including ecosystem-disturbance feedback), and mussel bed state transitions, respectively. The shape of compartment symbols differentiates mussel bed states (circles) from sequestered matter E (squares, mass per unit area). Ecosystem-disturbance feedback acts at the local scale (**a**) but is also implemented in the presence of abiotic regional coupling. Adapted from [87]

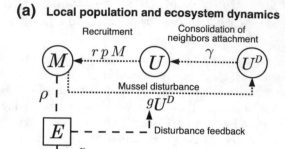

(a) Local population and ecosystem dynamics

M : mussel-covered area

U : empty area, surrounding mussel aggregates have consolidated their attachment and are not vulnerable to disturbance propagation

U^D : empty area from which disturbance can propagate as byssal threads at the interface with surrounding mussel aggregates have been damaged

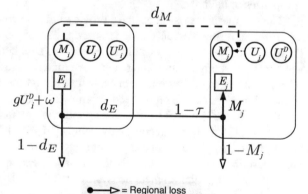

(b) Meta-ecosystem: propagule dispersal and regional matter transport

loss of sediments through physical processes (currents) other than disturbances. The system of equations describing the dynamics of this local ecosystem is (Fig. 4.9a):

$$\frac{dM}{dt} = \alpha\, p(M)\, UM - g(E, M)U^D M, \qquad (4.12a)$$

$$\frac{dU^D}{dt} = g(E, M)U^D M - \gamma U^D, \qquad (4.12b)$$

$$\frac{dE}{dt} = \rho M - E \cdot (g(E, M)U^D + \omega), \qquad (4.12c)$$

$$\{M, U, U^D\} \in [0, 1];\; M + U + U^D = 1.$$

We can first assume the rate of mussel bed disturbance $g(E, M)$ is independent of the amount of sequestered matter and is a constant parameter β ($g(E, M) = \beta$). We can also relax this assumption, and implement a feedback between the amount of matter sequestration and the rate of disturbance propagation. Based on field observations reported by [87], we define the rate of disturbance propagation, g, as a saturating function of the mass of sequestered matter per mussel-covered area (E_d):

$$g(E, M) = \beta \cdot \frac{a\, E_d + kh}{h + E_d}$$

$$E_d = \frac{E}{M N^2}$$

(4.13)

where N^2 is the maximum mussel cover, $a\beta$ and $k\beta$ are maximum and minimum disturbance rates, and h is half-saturation level of matter density.

Finally, scaling up this local ecosystem model to meta-ecosystem dynamics, is achieved by considering transport of organisms and matter (biotic and abiotic spatial coupling) between two homogeneous ecosystems, patches i and j (Fig 3.13b). First, we implement movement of organisms by rewriting the larval recruitment function p in patch i as a Poisson process:

$$p_i(M_i, M_j) = 1 - e^{-(f(1-d_M)M_i + f d_M M_j)},$$

$$i, j \in [1, 2]; i \neq j$$

(4.14)

where biotic connectivity (d_M) describes the proportion of larvae exported from each local ecosystem (Fig. 4.9b). Recruitment is defined by f, which aggregates per capita fecundity and larval transport.

At the meta-ecosystem level, resuspended matter following mussel disturbance can either be (i) sequestered by mussels in the recipient patch, or (ii) lost from the meta-ecosystem (Fig. 4.9b). We define abiotic connectivity (d_E) as the proportion of resuspended matter available for sequestration in the recipient patch. We assume all resuspended matter is exported from the disturbed ecosystem (no retention). Dynamics of matter in patch i within the meta-ecosystem then become:

$$\frac{dE_i}{dt} = \rho M_i - E_i \cdot [g(E_i, M_i)U_i^D + \omega] + d_E E_j \cdot [g(E_j, M_j)U_j^D + \omega] \quad (4.15)$$

Reference [87] analysed their model numerically starting from the observation that even in the absence of any effect of matter on mussel dynamics, mussel aggregation and the propagation of wave disturbances can induce self-sustained relaxation (non-sinusoidal) oscillations that result in rare and extreme events of matter resuspension. The non-resource effect of sequestered matter on mussel dynamics can further destabilize local dynamics through a positive ecosystem-disturbance feedback between sequestered matter and disturbance propagation.

At the meta-ecosystem level, the non-resource effect of matter on mussels through the ecosystem-disturbance feedback allows abiotic connectivity to control both local dynamics and spatial synchrony (see 3.4.1.1). These interacting local and regional effects of matter on dynamics can cause strong temporal variability in the retention of matter and low regional mussel cover under strong abiotic connectivity. As a result, the local ecosystem-disturbance feedback creates a strong positive effect of abiotic coupling on the retention (Fig. 4.10a) and average stocks (Fig. 4.10b) of matter (sediments in this case). It more specifically leads to a decreasing regional loss of matter as a function of abiotic coupling, compared to constant loss in the absence of feedback (Fig. 4.10a). The feedback also contributes to an increasing regional stock of matter with increasing abiotic connectivity, well above those resulting from

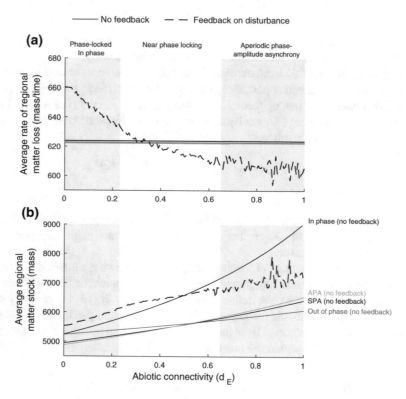

Fig. 4.10 Effect of abiotic connectivity (transport of matter) on dynamical regimes with and without ecosystem-disturbance feedback. Panels show long-term averages of **a** the rate of loss of matter from the meta-ecosystem and **b** the regional stock. In the absence of an ecosystem-disturbance feedback (solid lines) colours indicate the spatial synchrony regime controlled by biotic coupling: out-of-phase synchrony, symmetrical phase-amplitude asynchrony (SPA), in-phase synchrony, and asymmetric phase-amplitude asynchrony (APA). These regimes corresponds to Fig 3.13a–d respectively. In the presence of an ecosystem-disturbance feedback (dashed line), shaded areas delineate spatial synchrony regimes driven by abiotic coupling. Note that Near Phase Locking (NPL) is a special case of symmetrical phase-amplitude asynchrony where local mussel cover oscillations of both patches overtake each other in alternation, with extremely small fluctuations of the frequency and amplitude of local dynamics. Adapted from [87]

synchrony regimes observed in the absence of feedback, from in-phase, to out-of-phase and asynchronous dynamics (Fig. 4.10b; see also Sect. 3.4.1.1).

In the presence of a local feedback between sequestered matter and disturbance, the regional retention and average stock of matter strongly increase with abiotic coupling. In the absence of this feedback, only the average stock responds to abiotic connectivity. We can see from Fig. 4.10b that the in-phase regime (solid black line) leads to a strong increase in the average regional stock with abiotic coupling. This is interesting because it suggests a trade-off between the accumulation of matter, which is an important ecosystem function, and the persistence of the mussel population that would be negatively impacted by large in-phase (spatially-homogeneous) oscillations. Meta-ecosystem theories integrating non-resource effects can thus provide a new understanding of how foundations species can affect ecosystem functions across scales: processes that involve local non-resource effects of matter on disturbance and that promote abrupt and large amplitude changes are also predicted to scale up through abiotic coupling and facilitate the retention of matter over meta-ecosystems.

4.5.2 Matter as Information: Thresholds and Gradients Altering Movement

Throughout this book, we have made a key assumption that many a reader would pass over with little comment: organisms move in a simple, diffusive manner between ecosystems of the same type. While a useful approximation in many cases, organisms as simple as *E. coli* are capable of directed movements towards their resources [140]. In order to do so, organisms utilize the information that is available to them, which in many cases are gradients of chemicals and the resulting movement is known as chemotaxis [140].

Chemotaxis has been a topic of great interest to theoretical and mathematical ecologists for over half a century [140]. The standard chemotaxis model, called the Keller-Seger or PKS (after Patlak, Keller and Seger) model, is a system of two partial differential equations describing the dynamics of a population, $u(\mathbf{x}, t)$, and its chemoattractant, $v(\mathbf{x}, t)$:

$$\frac{\partial u}{\partial t} = \nabla \cdot (D_u(u, v)\nabla u - u\chi(u, v)\nabla v) + f(u, v) \tag{4.16a}$$

$$\frac{\partial v}{\partial t} = D_v\nabla^2 v + g(u, v) \tag{4.16b}$$

where $\mathbf{x}$ is the spatial dimension of the system (i.e. $\mathbf{x} \in \mathbb{R}^n$), ∇u and ∇v are the gradients of u and v in space, $\nabla \cdot$ is the divergence of a scalar field, ∇^2 is the Laplacian (or the divergence of the gradient), $f(u, v)$ and $g(u, v)$ describe the local, instantaneous changes in population density and chemoattractant concentration, respectively, D_v is the diffusion coefficient of the chemoattractant, $D_u(u, v)$ is the population diffusion

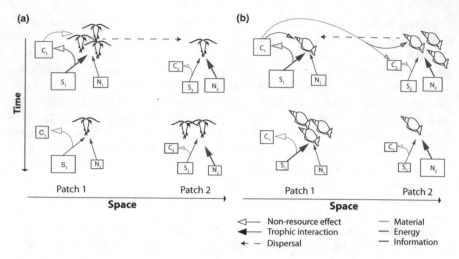

Fig. 4.11 Hypothetical benefits and costs of emitting a volatile organic compound C for a secreting algae S when competing against a non-secreting algae N in the presence of **a** a materially harmed herbivore (outlined copepod) or **b** an informationally attracted herbivore (outlined gastropod) in a two-patch system (1 and 2). **a** The volatile organic compound (VOC) is released by herbivory and harms the herbivore in high concentration (1), which could reduce its feeding on the algae and/or cause it to leave the patch, leading it to outcompete the non-secreting algae across the patches. **b** The VOC acts instead to attract the herbivore, causing increased herbivory in areas of high VOC and attracting herbivores from other patches, which leads to the secreting algae to be outcompeted regionally. Arrows same as Box 1 with the dashed black arrow indicates herbivore dispersal. Adapted from [118]

coefficient and $\chi(u, v)$ is the population chemotatic sensitivity coefficient [140]. The taxis part of the equation is $u\chi(u, v)\nabla v$ and it describes how individuals in the population move 'up' the chemoattractants concentration gradient (i.e. if they are in a place of low concentration, they will move towards an area with higher concentration). Similarly, if the substance was a chemorepellent, then organisms would move 'down' the gradient.

But are chemicals always attractants and repellents? In models such as these, in order to ensure analytical tractability, generally assume some sort of smooth function for chemotatic sensitivity. However, many examples in the behavioural and chemical ecology literature suggest that most chemicals only matter after a threshold concentration is reached [118, e.g.]. Furthermore, a chemical like a volatile organic compound that is a repellent to one organism may be an attractant to another (Fig. 4.11). The effects of these chemicals, which can be both material and informational, therefore requires an integrated approach with other material conditions that organisms are found within ecosystems.

At the moment, there are currently no meta-ecosystem models that have integrated the informational nature of matter, though there is growing interest to do so [105, 118]. Such work will require fundamental changes in the mathematical machinery of meta-ecosystems in order to approximate chemotatic responses in spatial fluxes of organisms. Furthermore, there are numerous dependencies on resource quality and ecological stoichiometry that drive the production of chemoattractants and repellents, requiring greater integration with ecological stoichiometry [118]. Finally, there are complex interactions between chemicals, which will necessitate consideration of multiple non-resource effects of matter to understand the informational landscape in meta-ecosystems [105, 118].

4.5.3   Meta-Ecosystems 2.1: Integration and Synthesis

Meta-ecosystems, as we have presented them, are highly idealized constructs. The local ecosystems generally have identical compartments, functions and parameters [54, 58, 87, 114–117], or small differences [57, 101]. Connectivity is defined at the level of the ecosystem and spatial fluxes occur through diffusive processes. Such simplifications allow us to use theorems to better characterize the dynamics and stability of our systems, but it comes at a cost.

Some of the most interesting ecological phenomena do not occur due to fluxes between similar ecosystems (e.g. lakes in a watershed), but between very different ecosystems [145]. Furthermore, the fluxes of organisms, energy and matter between ecosystems can differ not only in diffusive rates, but in the types of connections and the distance travelled [141]. Some fluxes may be unidirectional and others may exist according to the season [56]. In order to make progress on these issues, [56] called for a major update of the underlying theoretical constructs of meta-ecosystem ecology, moving from the version 1.0 presented here to a 2.0 version where it is more rooted in the field.

Here, we present a way forward towards this goal (Harvey et al. [70]). A potential framework we wish to use in this concluding section for meta-ecosystems can be expressed in a deceptively simple equation:

$$\frac{d\mathbf{x}}{dt} = \mathbf{F}(\mathbf{x}) + \mathbf{MCx} \tag{4.17}$$

where $\mathbf{x}$ is a $nm \times 1$ vector representing the amount of nutrient stock (or biomass) in each ecosystem compartment in each ecosystem arranged such that we list ecosystem compartment 1 in all ecosystems from ecosystem 1 to n, then ecosystem compartment 2 in all ecosystems from ecosystem 1 to n until we reach ecosystem compartment m in all ecosystems from ecosystem 1 to m, $\mathbf{F}(\mathbf{x})$ is a vector-valued function composed of nm functions $f_{i,k}(\mathbf{x_i})$ that describe the local dynamics of each ecosystem compartment in each ecosystem, where i is the ith ecosystem and k is the kth ecosystem compartment and $\mathbf{x_i} = (x_{i,1}, x_{i,2}, ..., x_{i,m})$, and $\mathbf{M}$ and $\mathbf{C}$ are $nm \times nm$ matrices describing the spatial flux rates of the compartments and their connectivity between ecosystems, respectively. Therefore, a great deal of complexity is hidden within this single equation that may be difficult to parse at first glance.

Fig. 4.12 Visualization of the multiplex meta-ecosystem conceptual schema with a simple three compartment (coloured circles), four ecosystem illustration. Black lines indicate flows of nutrients within the ecosystem, while the coloured lines indicate spatial flows. Note that the spatial flows can be uni- or bi-directional, in contrast with many other meta-ecosystem models

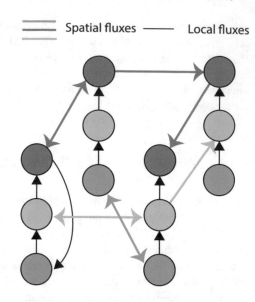

Our meta-ecosystem equation can be represented using multiplex networks (Fig. 4.12). Each ecosystem can have its own local flows between compartments (the connections between 'layers' can differ) and each compartment can have its own connectedness matrix (the connections within 'layers' do not need to match with other layers; Fig. 4.12). This approach differs from some other recent work that attempts to extend meta-ecosystem theory as it also allows for unidirectional flows, which are of great interest to field ecologists [125]. Therefore, it is possible in this framework to address the questions being asked of meta-ecosystem theory by empiricists, though it will likely be more complex to analyze.

Furthermore, it is possible to integrate within this general equation multiple ecological currencies as information and other non-resource effects can be added either to local interactions and/or spatial processes. For our example with the gastropods in Fig. 4.11, we could integrate the VOC as a given ecosystem compartment in all the patches, and its gradients would effect the spatial fluxes (and potentially the connectivity) of gastropods throughout the meta-ecosystem. Such effects are not currently possible to describe with earlier meta-ecosystem models, but Eq. 4.17 provides a potential way forward for integration.

Chapter 5
Conclusions

At the beginning of the twentieth century, the dominant organising concept of ecology was the biotic community, which privileged biotic interactions over all else in understanding ecological processes and dynamics [26, 89]. The soil, the climate and other abiotic components of the environment simply set the stage upon which the organisms act [26]. An alternative concept, the ecosystem, was developed to show the dependencies and relationships that exist between the inorganic and organic [164]. Over the next few decades, the ecosystem helped show that 'stuff' (i.e. nutrients, material) matters [143]. In a similar fashion, at the beginning of the twenty-first century, the metapopulation and metacommunity concepts are the predominant means by which ecologists think of the dynamic, spatial organization of ecological systems across multiple scales [65, 98]. The focus on the movement of organisms between habitats through dispersal has led to great advances, but once again the inorganic components of ecosystems only served as a background condition to be considered [110]. This situation led to the development of the meta-ecosystem concept and the emphasis on the fluxes of materials across ecosystem boundaries [110]. Throughout this text, we have attempted to show that spatial fluxes of 'stuff' also matters, which has important implications in both applied and theoretical ecology.

Contemporary conservation issues can benefit from this dynamic meta-ecosystem approach. Reducing the over-harvesting of natural resources (fisheries, forests) and the fragmentation of natural habitats, mitigating climate change impacts and making agriculture more sustainable can all be improved by looking across ecosystem boundaries. Addressing these environmental problems involve predicting the impacts of perturbations to the cycling and spatial flows of both living and non-living matter across scales. Towards that goal, challenges remain for improving the broad applicability of meta-ecosystem theories to natural systems, including the need for large ecological georeferenced datasets integrating biological and geochemical information. Much of this work is already being done within the complimentary framework

F. Guichard and J. Marleau, *Meta-Ecosystem Dynamics*, Lecture Notes
on Mathematical Modelling in the Life Sciences,
https://doi.org/10.1007/978-3-030-83454-8_5

of landscape ecology, which will require both synthesis and translation of concepts and results between the two fields.

The dynamic approach to meta-ecosystems we outlined in this book also provides opportunities to break away from equilibrium theories and to fully acknowledge the strong variability, both endogenous and exogenous, that characterizes most natural systems. Nonlinear ecological interactions and meta-ecosystem response to strong external disturbance events are central to the theories we covered here. This non-equilibrium meta-ecosystem theory shows how the cycling and spatial flows of matter are key components of nonlinear ecological dynamics that can, for example, contribute to spatially asynchronous oscillatory dynamics.

While meta-ecosystem dynamics has so far been developed mostly based on a classic ecosystem definition of matter as energy cycling between ecosystem compartments (e.g. from primary producers to herbivores), a more general meta-ecosystem theory can embrace the multiple roles of matter as resource, material and information with an explicit and dynamic conversion among these three currencies.

5.1   From Metacommunities to Meta-Ecosystem: Dynamic Feedbacks Between Diversity and Ecosystem Function

By developing an ecosystem theory building from metapopulation and metacommunity theories, we can address the dynamic nature of the relationship between community properties (e.g. diversity, relative abundance of species, food web topology) and ecosystem functions such as productivity or the retention and turnover of matter within ecosystem compartments. This relationship between biodiversity and ecosystem functions (BEF) has received much attention in recent year, boosted by the accelerating loss and reshuffling of species diversity worldwide and its documented impact on ecosystem functions human populations depend on. BEF has also been at the center of many controversies, in part explained by the need to establish clear statistical BEF relationships that can be communicated to the public, to managers and policy makers [108]. A a result, causality of BEF relationships has been assumed more than studied, with the loss of diversity being presented as the driver of ecosystem degradation. The research on BEF has also lacked integration of ecosystem-level functions, falling back on available proxy information such as biomass [44].

What meta-ecosystem dynamics offers in that context is a framework to understand feedbacks between biodiversity and ecosystem functions emerging from underlying ecosystem processes, including those involving flows of matter. For example, spatial flows of matter between ecosystems initiated by local species colonization can constitute a mechanism of coexistence, thus affecting diversity in turn feeding back on flows of matter that depend of the recycling of biomass. This feedback is predicted by meta-ecosystem models at equilibrium (Chap. 2), leading to apparent positive interactions among competing species in spatially heterogeneous environments [57] or extending conditions for coexistence in spatially homogeneous environments [114]. It is also observed under non-equilibrium dynamics (Chap. 3) where predator-prey interactions and the recycling of there biomass interact to cause oscillatory dynamics that promote their persistence and overall ecosystem productivity [115]. Similar feedbacks can emerge from more a detailed consideration of ecological stoichiometry [117] and of the topology of spatial networks [116]. The message here is simple, yet important for conservation: species can, through species interactions, initiate ecosystem functions that promote their persistence and the assembly of complex communities. Focusing on the loss of biodiversity as a cause of ecosystem health degradation might misleadingly emphasize ecosystem function as a service resulting from biodiversity.

5.2 Applying Meta-Ecosystem Theories to Natural Ecosystems

Meta-ecosystem theories were developed as mathematical models that could extend well-known metapopulation and metacommunity theories to ecosystem dynamics. But while metapopulation theories came early enough to be adopted by experimental and field ecologists, the meta-ecosystem framework, as an extension to metapopulation theories, remained somewhat disconnected from a long tradition of ecosystem studies integrating spatial fluxes of matter across ecosystem boundaries [170]. The relevance of the meta-ecosystem framework remains its relative simplicity and tractability that have lead to a body of predictions offering a new synthesis between population, community and ecosystem processes. Microcosm [68] and field [96] experiments have recently taken advantage of that simplicity to test meta-ecosystem predictions. Despite these advances, some key properties of natural systems are still missing from most meta-ecosystem models that would greatly improve their applicability to natural systems [56].

The challenge of improving the applicability of meta-ecosystem theories is certainly worth the effort. It is first one way to achieve a synthesis of community and ecosystem ecology by providing a simple formalism to tackle high-dimensional systems combining the dynamics of many species and of fluxes of matter. Landscape ecology is an other strong candidate to achieve this synthesis, with its emphasis on complex spatial structure and environmental heterogeneity characterizing natural systems [170]. It is no surprise then that most recent discussions of what future holds for meta-ecosystem theory converge at the interface of simple meta-ecosystem dynamical models and of landscape ecological approaches to natural systems [56, 102, 141].

Furthermore, the strive for greater realism in meta-ecosystem theory will require better integration with both biogeochemistry [159] and ecosystem ecology [22]. Currently, the key processes of nutrient recycling and decomposition are highly simplified in meta-ecosystem theory, which belies the complex interactions that exist between the quality of dead organic matter and its ability to be decomposed (e.g. [17,

113]). The rich body of theory and modelling of these processes (e.g. [38]) is currently being integrated into food web models (e.g. [16, 24]), suggesting that an integration with meta-ecosystem theory would be highly fruitful.

5.3 A Non-equilibrium Meta-Ecosystem Theory of Ecosystem Management

A non-equilibrium theory of spatial ecological dynamics certainly needs to consider the recycling and movement of organic and inorganic matter. The theoretical framework of meta-ecosystems predicts ecological consequences of limited flows of organic and inorganic matter across space. This emerging ecological theory [110] and its application to natural systems will provide a good case study to build and validate a theory of non-equilibrium spatial dynamics. For example, recent theory suggests flows of organic and inorganic matter can deeply transform the structure and stability of communities (Chap. 2), and generate nonlinear feedbacks and explain spatiotemporal heterogeneity (Chap. 3). These theoretical studies echo recent calls for whole-ecosystem management of natural resources that are embedded in complex interaction networks and flows of (in)organic matter. Ecosystem-based management is thus an integrative approach that can address the sustainability of ecosystem services across scales.

As an example, coastal ecosystems are considered among the most productive and the most threatened by human activities and climate change [129], and are thus prime candidates for ecosystem-based management approaches [126]. They show strong fluctuations in the distribution of species and in ecosystem functions (e.g. nutrient recycling and primary production). This is in part why ecologists and managers have called for ecosystem-based management of coastal ecosystems. However, the prevailing ecological paradigm informing management is based on an equilibrium theory of community and ecosystem structure. This theory predicts the equilibrium and fast adjustment of marine ecosystems to oceanographic change at the community level through species interactions [15, 27, 128], or at the species level through trait-based shifts in species range [42].

Despite empirical evidence, current theories of coastal (and other) ecosystems assume that nutrients are well mixed across large spatial scales. These simplifying assumptions contained within ecological theories are in sharp contrast with our detailed understanding of transport and cycles of nutrient and organic matters in oceans. Coastal systems strongly depend on coupling between benthic and pelagic compartments [32, 127], and spatial fluxes are therefore controlled across a very broad range of spatial and temporal scales [146], from fast advective mixing and stratification [7] to storage over geological time scales. Given the accumulating evidence of strong meta-ecosystem dynamics in coastal ecosystems, one task is to test for the existence and implications of nonlinear feedbacks over regional scales that are targeted by managers and policy makers. The sudden shifts in the state of regional fisheries documented over the last decades [41] certainly suggest the relevance and urgency of this task.

Empirical evidence of coupling between ecosystem-level (e.g. nutrient and oxygen fluxes) and community-level (e.g. population growth, predation) processes has emerged in the recent literature [96, 142]. These studies conducted over local scales are compatible with the regional regulation of coastal ecosystems by limited spatial fluxes of (in)organic matter, but the role of such spatial fluxes has yet to be integrated within general theories of coastal ecosystems. At the ecosystem level, most coastal ecosystems are limited by nitrogen [10], which is affected by human activities (e.g. agricultural runoffs). Nutrient cycling through ecosystems and their spatial flows in the water column and across benthic and pelagic compartments has been documented [49] and built into predictive models by oceanographers [10]. This body of literature still has to be translated into a general ecological theory. More specifically, we still need to understand how the transport of in(organic) matter interacts with ecological processes and dispersal to affect the stability of communities and ecosystem functions.

5.4 Integrating Multiple Currencies of Matter to Meta-Ecosystem Dynamics

We have briefly presented a meta-ecosystem framework that shows how recognizing the energetic, material and informational facets of matter can improve our ecological understanding of its effects in ecosystems [118]. Quantifying these currencies as stocks of matter helps us integrate between successful resource-based theories like ecological stoichiometry with non-resource ideas such as ecological engineering. Bridging this gap can only be done if we talk about and use the same currencies, and the metrics used to quantify them; thus, integrative theory needs to be able to go from mole to joule to bit and back again. Clearly, much work is left to be done, but we hope this work helps spur deeper integration across theoretical frameworks in ecology, with the goal of making predictions that apply to human-impacted ecosystems.

Humans are changing the information in the environment and this is altering species interactions and the flows of matter within and among ecosystems. Integrating meta-ecosystem dynamics across multiple currencies of matter enables changes in environmental conditions in one ecosystem to be linked with impacts in other ecosystems, which is common in human-impacted systems. For example, changes in pH due to agricultural practices and/or types of crops in terrestrial landscapes can increase the availability and facilitate the runoff of nutrients such as phosphorus and nitrogen [40]. These non-local impacts of matter can also occur when humans disrupt the ecosystems that produce the matter. The decline of kelp beds across the globe reduces the raw material that creates habitat well beyond the boundaries of the kelp bed [94]. The introduction of exotic species creates new informational landscapes, which can disrupt the local trophic interactions [21]. Both these impacts can be readily derived when the material and informational impacts of organisms are combined in analyses but are otherwise difficult to incorporate in a resource-based analysis.

5.5 Concluding Remarks

Meta-ecosystem dynamics borrows from and builds on both metapopulation and ecosystem theories. By doing so, it provides an integrated framework to study flows of matter, of living and non-living ecosystem compartments, and through multiple processes including movement and trophic interactions. As a mechanistic theory, it can predict relationships between properties of populations, communities and whole ecosystems from dynamic processes across these levels of organization. As a theory of spatial dynamics, it formulates fluxes of matter as the common currency to study these properties across scales. The abstract and simplistic nature of most meta-ecosystem models also make them tractable despite the complexity of processes they tackle, which facilitates the development of a general and inclusive body of theory. However, as a relatively recent concept, meta-ecosystem dynamics runs the risk of overlooking decades of contributions to spatial dynamics of ecosystems. As an abstract and simplistic representation of natural systems, meta-ecosystem theories are still lacking a deep connection with the complex spatial structure characterizing natural landscapes and of highly heterogeneous movement patterns of individuals and matter. As meta-ecosystem theory becomes increasingly applicable to natural systems, it will pass the real test of its relevance if it offers novel insights that retain the generality of simple models and that add to decades of empirical research in ecosystem sciences. We hope the models we have outlined in this book can contribute to bridging the gap between this rich literature and an understanding of ecosystems that stemmed from theoretical ecology.

Glossary

Ecology has over time adopted a very diverse and heterogeneous terminology to define key processes and quantities. This variability reflects the history of the field, but also the importance of setting these processes and quantities in their biological context. As a result, processes such as diffusion can be referred to by multiple terms depending on the nature of the diffusing variable despite being implemented using the same equation. We have made an effort to adopt a standard terminology while recognizing the ecological relevance of a context-specific terminology. What follows is a list of main ecological processes and quantities that used in this book. For each we provide a working definition in the context of meta-ecosystem models and specific ecological context where other terms can be used instead.

Connectivity Displacement of a portion of an ecosystem compartment between spatially distinct ecosystems. Throughout the book, we assess connectivity as the combination of connectedness characterizing physical properties of landscapes allowing for potential fluxes, and of spatial fluxes characterizing the potential rate of displacement of specific compartments. Spatial fluxes can be driven by the active or passive movement of individual organisms, or by the transport of organic or inorganic matter. Diffusion provides a description of passive movement and transport. Dispersal is used to refer to movement of biotic compartments via propagules such as seeds or larvae. The partitioning of connectivity into connectedness and spatial fluxes is derived in Chap. chpt3 and more explicitly in Chap. chpt4. While connectivity defines displacements within a meta-ecosystem, we also define inputs and outputs that are exchanges with locations that are external to an open meta-ecosystem. No such exchanges are allowed in closed meta-ecosystems.

Consumer Heterotrophic (biotic) ecosystem compartment. Consumers can be used to refer to any living ecosystem compartment other than autotrophs (also called primary producers). Consumers feeding on primary producers are herbivores. Consumers feeding on other consumers are also called predators, in which case the consumer's resource is called a prey.

Resource Any ecosystem compartment used as a source of energy by another compartment. In ecosystem models that term is often reserved for abiotic compartments such as nutrients while community ecologists often use resource to refer to primary producers.

Patch Spatially distinct and (usually homogeneous) ecosystem. In meta-ecosystem models, a patch corresponds to a single (local) ecosystem connected by spatial movement to other ecosystems to for a meta-ecosystem. Patches are sometimes called habitats or simply ecosystems.

Ecosystem A community of organisms interacting with its abiotic environment through the fluxes of matter and energy.

Ecosystem compartment Individual state variable within a local ecosystem. A compartment corresponds to a distinct form matter, living, organic or inorganic. A compartment can describe the stock of a particular element, the mass of organic matter such as detritus, the biomass or abundance of a species. Ecosystem compartments can be connected by trophic (i.e. feeding) or non-trophic interactions defining the transformation of matter through time.

Ecosystem function Very broadly referring to the production of quantities associated with individual or aggregated compartments that is considered important for the persistence or resilience of the (meta-)ecosystem. These quantities can be equilibrium size or turnover rate of matter in, or between compartments, within or between ecosystems, and within and through the meta-ecosystem. This description is compatible with ecosystem function corresponding to ecological processes (e.g. recycling) or various measures of ecosystem structure (e.g. set of interactions among compartments).

Equilibrium versus non-equilibrium dynamics Equilibrium refers to long-term dynamics characterized by a steady (constant) value of all state variable(s). Non-equilibrium dynamics is used to refer to dynamics that does not display long-term constant values of its state variables. For example, the logistic equation is a differential equation for the dynamics of a single population and its state variable, population size, has an equilibrium positive value (the so-called carrying capacity). We typically talk about equilibrium dynamics when the equilibrium point is locally stable (see Stability). Non-equilibrium dynamics can then be used to describe dynamics that does not possess or reach such an equilibrium value. In most ecological contexts, non-equilibrium dynamics can result from the lack of any stable equilibrium and by the presence instead of stable oscillations. Non-equilibrium dynamics can also be caused by external (environmental) disturbances keeping dynamics away from an otherwise stable equilibrium. Following a disturbance (perturbation), a dynamic system can reach its stable equilibrium after a transient time during which changes in the state variables are observed. Repeated disturbances can thus prevent dynamic systems from ever reaching its stable equilibrium thus resulting in non-equilibrium dynamics.

Stability In most cases in this book, stability is short for local stability and is used to describe dynamic regimes of models. It more specifically characterizes the ability of a dynamic system to reach a particular dynamic regime from states that are in close proximity to that regime. Stability can apply to any dynamic regimes such as equilibrium points and oscillations (orbits). A system is said to be locally stable with reference to a dynamic regime if it reaches that regime following a small perturbation away from that regime. We will sometimes use other definitions of stability used in ecology: Stability can be a quantity inversely proportional to the amount of variability in state variables (variance or coefficient of variation), or it could be associated with the concept of resilience which, in ecology, usually corresponds to the rate of return of a system to its stable regime following a perturbation.

Spatial synchrony In ecology, synchrony has been used very broadly to refer to the degree of association between variables, with a sometimes implicit focus on the coherence of variations and of local maxima in addition to their overall correlation. Spatial synchrony is used to describe such association between geographically distinct time series, typically of the same variable. Ecologists have also used spatial synchrony within the more restricted context of coupled oscillators where is refers to the phase (and sometimes amplitude) difference between time series containing periodic components. This is the definition we adopt in this book. Within this context, the phase is defined over a 2π cycle and phase synchrony refers to an equilibrium phase difference between locations. Special cases of phase synchrony include in-phase (zero difference) and out-of-phase (π phase difference). Phase-locked synchrony is often used to refer to other equilibrium states. Phase asynchrony refers to deviation from phase synchrony where non-steady phase difference is observed associated with heterogeneous frequencies of oscillations between locations. It should be noted that synchrony is sometimes (but not in this book) reserved for the specific case of in-phase synchrony, with asynchrony then used for any deviation from in-phase synchrony.

References

1. Abbott, K.C.: A dispersal-induced paradox: synchrony and stability in stochastic metapopulations. Ecol. Lett. **14**(11), 1158–1169 (2011). https://doi.org/10.1111/j.1461-0248.2011.01670.x
2. Allen, D.C., Wesner, J.S.: Synthesis: comparing effects of resource and consumer fluxes into recipient food webs using meta-analysis. Ecology **97**(3), 594–604 (2016). https://doi.org/10.1890/15-1109.1
3. Allstadt, A.J., Liebhold, A.M., Johnson, D.M., Davis, R.E., Haynes, K.J.: Temporal variation in the synchrony of weather and its consequences for spatiotemporal population dynamics. Ecology **96**(11), 2935–2946 (2015). https://doi.org/10.1890/14-1497.1
4. Angelini, C., Altieri, A.H., Silliman, B.R., Bertness, M.D.: Interactions among foundation species and their consequences for community organization, biodiversity, and conservation. BioScience **61**(10), 782–789 (2011). https://doi.org/10.1525/bio.2011.61.10.8
5. Armstrong, R.A., McGehee, R.: Competitive exclusion. Amer. Naturalist 115(2), 151–170 (1980). https://doi.org/10.1086/283553
6. Ashwin, P., King, G.P., Swift, J.W.: Three identical oscillators with symmetric coupling. Nonlinearity **3**(3), 585–601 (1990). https://doi.org/10.1088/0951-7715/3/3/003
7. Barth, J.A., Menge, B.A., Lubchenco, J., Chan, F., Bane, J.M., Kirincich, A.R., McManus, M.A., Nielsen, K.J., Pierce, S.D., Washburn, L.: Delayed upwelling alters nearshore coastal ocean ecosystems in the northern California current. Proceedings of the National Academy of Sciences of the United States of America 104(10), 3719–3724 (2007). https://doi.org/10.1073/pnas.0700462104
8. Belykh, I., Piccardi, C., Rinaldi, S.: Synchrony in tritrophic food chain metacommunities. J. Biolog. Dyn. **3**(5), 497–514 (2009)
9. Benke, A.C.: Secondary production as part of bioenergetic theory–contributions from freshwater benthic science. River Res. Appl. **26**(1), 36–44 (2010). https://doi.org/10.1002/rra.1290
10. Billen G, Lancelot, C.: Modelling benthic nitrogen cycling in temperate coastal ecosystems. In: Blackburn T, Sorensen J (eds.) Nitrogen Cycling in Coastal Marine Environments, Wiley, chap 14, pp. 341–378 (1988)
11. Bjornstad, O.N.: Trading space for time in population dynamics. Trends Ecol. Evolut. **16**(3), 124 (2001)
12. Bjørnstad, O.N., Ims, R.A., Lambin, X.: Spatial population dynamics: Analyzing patterns and processes of population synchrony. Trends Ecol. Evolut. 14(11), 427–432 (1999). https://doi.org/10.1016/S0169-5347(99)01677-8

F. Guichard and J. Marleau, *Meta-Ecosystem Dynamics*, Lecture Notes on Mathematical Modelling in the Life Sciences, https://doi.org/10.1007/978-3-030-83454-8

13. Blasius, B., Huppert, A., Stone, L.: Complex dynamics and phase synchronization in spatially extended ecological systems. Nature **399**(6734), 354–359 (1999)
14. Briggs, C.J., Hoopes, M.F.: Stabilizing effects in spatial parasitoid-host and predator-prey models: a review. Theoret. Popul. Biol. **65**(3), 299–315 (2004)
15. Broitman, B.R., Blanchette, C.A., Menge, B.A., Lubchenco, J., Krenz, C., Foley, M., Raimondi, P.T., Lohse, D., Gaines, S.D.: Spatial and temporal patterns of invertebrate recruitment along the West Coast of the United States. Ecolog. Monogr. **78**(3), 403–421 (2008)
16. Buchkowski, R.W., Leroux, S.J., Schmitz, O.J.: Microbial and animal nutrient limitation change the distribution of nitrogen within coupled green and brown food chains. Ecology **100(5):e02,674**,(2019a). https://doi.org/10.1002/ecy.2674
17. Buchkowski, R.W., Schmitz, O.J., Bradford, M.A.: Nitrogen recycling in coupled green and brown food webs: Weak effects of herbivory and detritivory when nitrogen passes through soil. J. Ecol. **107**(2), 963–976 (2019b). https://doi.org/10.1111/1365-2745.13079
18. Cantrell, R.S., Cosner, C.: Spatial Ecology via Reaction-Diffusion Equations. Wiley (2004)
19. Cazelles, B., Bottani, S., Stone, L.: Unexpected coherence and conservation. Proc. Royal Soc. London Ser. B-Biol. Sci. **268**(1485), 2595–2602 (2001)
20. Cazelles, B., Chavez, M., McMichael, A.J., Hales, S.: Nonstationary Influence of El Niño on the Synchronous Dengue Epidemics in Thailand. PLoS Medicine **2**(4), 313–318 (2005). https://doi.org/10.1371/journal.pmed.0020106
21. Chabaane, Y., Laplanche, D., Turlings, T.C.J., Desurmont, G.A.: Impact of exotic insect herbivores on native tritrophic interactions: A case study of the African cotton leafworm, Spodoptera littoralis and insects associated with the field mustard Brassica rapa. J. Ecol. **103**(1), 109–117 (2015). https://doi.org/10.1111/1365-2745.12304
22. Chapin, F.S.I., Matson, P.A., Vitousek, P.M.: Principles of Terrestrial Ecosystem Ecology, 2nd edn. Springer, New York (2011)
23. Chave, J.: Neutral theory and community ecology. Ecology Letters 7, 241–253 (2004). https://doi.org/10.1111/j.1461-0248.2003.00566.x
24. Cherif, M., Loreau, M.: Plant-herbivore-decomposer stoichiometric mismatches and nutrient cycling in ecosystems. Presented at the (2013). https://doi.org/10.1098/rspb.2012.2453
25. Chesson, P.: Mechanisms of maintenance of species diversity. Ann. Rev. Ecol. Systematics 31, 343 (2000)
26. Clements, F.E.F.E.: Plant Succession; an Analysis of the Development of Vegetation. Carnegie Institution of Washington, Washington (1916)
27. Connolly, S.R., Roughgarden, J.: Theory of marine communities: competition, predation, and recruitment-dependent interaction strength. Ecol. Monogr. **69**(3), 277–296 (1999)
28. Cooper, G.J.: The Science of the Struggle for Existence?: On the Foundations of Ecology. Cambridge Studies in Philosophy and Biology, Cambridge University Press, Cambridge, UK (2003)
29. Coulson, T., Rohani, P., Pascual, M.: Skeletons, noise and population growth: the end of an old debate? Trends Ecol. Evolut. **19**(7), 359–364 (2004)
30. Crain, C.M., Bertness, M.D.: Ecosystem engineering across environmental gradients: implications for conservation and management. BioScience **56**(3), 211–218 (2006). https://doi.org/10.1641/0006-3568(2006)056[0211:EEAEGI]2.0.CO;2
31. Cuddington, K., Wilson, W.G., Hastings, A.: Ecosystem engineers: feedback and population dynamics. Amer. Naturalist **173**(4), 488–498 (2009). https://doi.org/10.1086/597216
32. Dame, R., Dankers, N., Prins, T., Jongsma, H., Smaal, A.: The influence of mussel beds on nutrients in the western wadden sea and eastern scheldt estuaries. Estuaries **14**(2), 130–138 (1991)
33. Daufresne, T., Hedin, L.: Plant coexistence depends on ecosystem nutrient cycles: Extension of the resource-ratio theory. Presented at the (2005). https://doi.org/10.1073/pnas.0406427102
34. Daufresne, T., Loreau, M.: Plant-herbivore interactions and ecological stoichiometry: when do herbivores determine plant nutrient limitation? Ecol. Lett. **4**, 196–206 (2001)
35. DeAngelis, D.L.: General concepts of nutrient flux and stability. In: DeAngelis, D.L. (eds.) Dynamics of Nutrient Cycling and Food Webs, Population and Community Biology Series,

pp. 17–37. Springer, Netherlands, Dordrecht (1992a). https://doi.org/10.1007/978-94-011-2342-6_2

36. DeAngelis, D.L.: Herbivores and nutrient recycling. In: DeAngelis, D.L. (ed.) Dynamics of Nutrient Cycling and Food Webs, Population and Community Biology Series, pp. 108–122. Springer, Netherlands, Dordrecht (1992b). https://doi.org/10.1007/978-94-011-2342-6_6

37. DeAngelis, D.L.: Introduction. In: DeAngelis, D.L. (ed.) Dynamics of Nutrient Cycling and Food Webs, Population and Community Biology Series, Springer Netherlands, Dordrecht, pp. 1–16 (1992c). https://doi.org/10.1007/978-94-011-2342-6_1

38. DeAngelis, D.L.: Nutrient interactions of detritus and decomposers. In: DeAngelis, D.L. (ed.) Dynamics of Nutrient Cycling and Food Webs, Population and Community Biology Series, pp. 123–141. Springer, Netherlands, Dordrecht (1992d). https://doi.org/10.1007/978-94-011-2342-6_7

39. DeAngelis, D.L., Waterhouse, J.: Equilibrium and nonequilibrium concepts in ecological models. Ecol. Monogr. **57**(1), 1–21 (1987)

40. Devau, N., Le Cadre, E., Hinsinger, P., Gérard, F.: A mechanistic model for understanding root-induced chemical changes controlling phosphorus availability. Ann. Botany **105**(7), 1183–1197 (2010). https://doi.org/10.1093/aob/mcq098

41. deYoung, B., Barange, M., Beaugrand, G., Harris, R., Perry, R.I., Scheffer, M., Werner, F.: Regime shifts in marine ecosystems: detection, prediction and management. Trends Ecol. Evolut. **23**(7), 402–409 (2008). https://doi.org/10.1016/j.tree.2008.03.008

42. Donelson, J.M., Sunday, J.M., Figueira, W.F., Gaitán-Espitia, J.D., Hobday, A.J., Johnson, C.R., Leis, J.M., Ling, S.D., Marshall, D., Pandolfi, J.M., Pecl, G., Rodgers, G.G., Booth, D.J., Munday, P.L.: Understanding interactions between plasticity, adaptation and range shifts in response to marine environmental change. Presented at the (2019). https://doi.org/10.1098/rstb.2018.0186

43. Edwards, C.A., Powell, T.A., Batchelder, H.P.: The stability of an NPZ model subject to realistic levels of vertical mixing. J. Mar. Res. **58**(1), 37–60 (2000). https://doi.org/10.1357/002224000321511197

44. Eisenhauer, N., Schielzeth, H., Barnes, A.D., Barry, K.E., Bonn, A., Brose, U., Bruelheide, H., Buchmann, N., Buscot, F., Ebeling, A., Ferlian, O., Freschet, G.T., Giling, D.P., Hättenschwiler, S., Hillebrand, H., Hines, J., Isbell, F., Koller-France, E., König-Ries, B., de Kroon, H., Meyer, S.T., Milcu, A., Müller, J., Nock, C.A., Petermann, J.S., Roscher, C., Scherber, C., Scherer-Lorenzen, M., Schmid, B., Schnitzer, S.A., Schuldt, A., Tscharntke, T., Türke, M., van Dam, N.M., van der Plas, F., Vogel, A., Wagg, C., Wardle, D.A., Weigelt, A., Weisser, W.W., Wirth, C., Jochum, M.: A multitrophic perspective on biodiversity-ecosystem functioning research. Presented at the (2019). https://doi.org/10.1016/bs.aecr.2019.06.001

45. Ellner, S.P., McCauley, E., Kendall, B.E., Briggs, C.J., Hosseini, P.R., Wood, S.N., Janssen, A., Sabelis, M.W., Turchin, P., Nisbet, R.M., Murdoch, W.W.: Habitat structure and population persistence in an experimental community. Nature **412**(6846), 538–543 (2001)

46. Feng, W., Hinson, J.: Stability and pattern in two-patch predator-prey population dynamics. Discrete Contin. Dyn. Syst. Suppl. **2005**, 268–279 (2005)

47. Feng, W., Rock, B., Hinson, J.: On a new model of two-patch predator-prey system with migration of both species. J. Appl. Anal. Comput. **1**(2), 193–203 (2011)

48. Fox, J.W.: Interpreting the 'selection effect' of biodiversity on ecosystem function. Ecol. Lett. **8**(8), 846–856 (2005). https://doi.org/10.1111/j.1461-0248.2005.00795.x

49. Frechette, M., Butman, C.A., Geyer, W.R.: The importance of boundary-layer flows in supplying phytoplankton to the benthic suspension feeder, mytilus edulis l. Limnology Oceanogr. **34**(1), 19–36 (1989)

50. Goldwyn, E.E., Hastings, A.: When can dispersal synchronize populations? Theoret. Popul. Biol. **73**(3), 395–402 (2008)

51. Gonzalez, A., Germain, R.M., Srivastava, D.S., Filotas, E., Dee, L.E., Gravel, D., Thompson, P.L., Isbell, F., Wang, S., Kéfi, S., Montoya, J., Zelnik, Y.R., Loreau, M.: Scaling-up biodiversity-ecosystem functioning research. Ecol. Lett. **23**(4), 757–776 (2020). https://doi.org/10.1111/ele.13456

52. Gouhier, T.C., Guichard, F., Menge, B.A.: Ecological processes can synchronize marine population dynamics over continental scales. Proc. Natl Acad. Sci. USA **107**(18), 8281–8286 (2010). https://doi.org/10.1073/pnas.0914588107

53. Gouhier, T.C., Guichard, F., Menge, B.A.: Designing effective reserve networks for nonequilibrium metacommunities. Ecological Applications **23**(6), 1488–1503 (2013). https://doi.org/10.1890/12-1801.1

54. Gounand, I., Mouquet, N., Canard, E., Guichard, F., Hauzy, C., Gravel, D.: The paradox of enrichment in Metaecosystems. Amer. Naturalist **184**(6), 752–763 (2014). https://doi.org/10.1086/678406

55. Gounand, I., Harvey, E., Ganesanandamoorthy, P., Altermatt, F.: Subsidies mediate interactions between communities across space. Oikos **126**(7), 972–979 (2017). https://doi.org/10.1111/oik.03922

56. Gounand, I., Harvey, E., Little, C.J., Altermatt, F.: Meta-ecosystems 2.0: rooting the theory into the field. Trends Ecol. Evolut. **33**(1), 36–46 (2018). https://doi.org/10.1016/j.tree.2017.10.006

57. Gravel, D., Guichard, F., Loreau, M., Mouquet, N.: Source and sink dynamics in metaecosystems. Ecology **91**(7), 2172–2184 (2010a). https://doi.org/10.1890/09-0843.1

58. Gravel, D., Mouquet, N., Loreau, M., Guichard, F.: Patch dynamics, persistence, and species coexistence in Metaecosystems. Amer. Naturalist **176**(3), 289–302 (2010b). https://doi.org/10.1086/655426

59. Grover, J.P., Holt, R.D.: Disentangling resource and apparent competition: realistic models for plant-herbivore communities. J. Theoret. Biol. **191**, 353–376 (1998)

60. Gruner, D.S., Smith, J.E., Seabloom, E.W., Sandin, S.A., Ngai, J.T., Hillebrand, H., Harpole, W.S., Elser, J.J., Cleland, E.E., Bracken, M.E.S., Borer, E.T., Bolker, B.M.: A cross-system synthesis of consumer and nutrient resource control on producer biomass. Ecol. Lett. **11**(7), 740–755 (2008). https://doi.org/10.1111/j.1461-0248.2008.01192.x

61. Guichard, F., Halpin, P.M., Allison, G.W., Lubchenco, J., Menge, B.A.: Mussel disturbance dynamics: signatures of oceanographic forcing from local interactions. Amer. Naturalist **161**(6), 889–904 (2003). https://doi.org/10.1086/375300

62. Guichard, F., Zhang, Y., Lutscher, F.: The emergence of phase asynchrony and frequency modulation in metacommunities. Theoret. Ecol. **12**(3), 329–343 (2019). https://doi.org/10.1007/s12080-018-0398-8

63. Gülzow, N., Wahlen, Y., Hillebrand, H.: Metaecosystem dynamics of marine phytoplankton alters resource use efficiency along stoichiometric gradients. Amer. Naturalist **193**(1), 35–50 (2018). https://doi.org/10.1086/700835

64. Gurney, W.S.C., Gurney, W.S.C., Nisbet, R.M.: PoTPERM, Nisbet: Ecological Dynamics. Oxford University Press (1998)

65. Hanski, I.: Metapopulation Ecology Oxford Series in Ecology and Evolution. Oxford University Press, Oxford (1999)

66. Harfoot, M.B.J., Newbold, T., Tittensor, D.P., Emmott, S., Hutton, J., Lyutsarev, V., Smith, M.J., Scharlemann, J.P.W., Purves, D.W.: Emergent global patterns of ecosystem structure and function from a mechanistic general ecosystem model. PLOS Biol. **12**(4), e1001, 841 (2014). https://doi.org/10.1371/journal.pbio.1001841

67. Harpole, W.S., Ngai, J.T., Cleland, E.E., Seabloom, E.W., Borer, E.T., Bracken, M.E., Elser, J.J., Gruner, D.S., Hillebrand, H., Shurin, J.B., Smith, J.E.: Nutrient co-limitation of primary producer communities. Ecol. Lett. **14**(9), 852–862 (2011). https://doi.org/10.1111/j.1461-0248.2011.01651.x

68. Harvey, E., Gounand, I., Ganesanandamoorthy, P., Altermatt, F.: Spatially cascading effect of perturbations in experimental meta-ecosystems. Presented at the (2016). https://doi.org/10.1098/rspb.2016.1496

69. Harvey, E., Gounand, I., Fronhofer, E.A., Altermatt, F.: Metaecosystem dynamics drive community composition in experimental, multi-layered spatial networks. Oikos **129**(3), 402–412 (2020). https://doi.org/10.1111/oik.07037

70. Harvey, E., Marleau, J., Gounand, I., et al.: A general meta-ecosystem model to predict ecosystem function at landscape extents. Authorea (2021). https://doi.org/10.22541/au.162799968.80128369/v1

71. Hassell, M.P., Comins, H.N., May, R.M.: Spatial structure and chaos in insect population-dynamics. Nature **353**(6341), 255–258 (1991)

72. Hassell, M.P., Comins, H.N., May, R.M.: Species coexistence and self-organizing spatial dynamics. Nature **370**(6487), 290–292 (1994)

73. Hastings, A.: Disturbance, coexistence, history, and competition for space. Theoret. Popul. Biol. **18**, 363–373 (1980)

74. Hastings, A., Higgins, K.: Persistence of transients in spatially structured ecological models. Science (New York, NY) **263**(5150), 1133–1136 (1994). https://doi.org/10.1126/science.263.5150.1133

75. He, D., Stone, L.: Spatio-temporal synchronization of recurrent epidemics. Proc. Royal Soc. London Ser. B: Biol. Sci. **270**(1523), 1519–1526 (2003). https://doi.org/10.1098/rspb.2003.2366

76. Henden, J.A., Ims, R.A., Yoccoz, N.G.: Nonstationary spatio-temporal small rodent dynamics: Evidence from long-term Norwegian fox bounty data. J. Animal Ecol. **78**(3), 636–645 (2009). https://doi.org/10.1111/j.1365-2656.2008.01510.x

77. Holland, M.D., Hastings, A.: Strong effect of dispersal network structure on ecological dynamics. Nature **456**(7223), 792–U76 (2008). https://doi.org/10.1038/nature07395

78. Holt, R.D., Bonsall, M.B.: Apparent competition. Ann. Rev. Ecol. Evolut. System. **48**(1), 447–471 (2017). https://doi.org/10.1146/annurev-ecolsys-110316-022628

79. Hoppensteadt, F.C., Izhikevech, E.M.: Weakly Connected Neural Networks. Springer, New York (1997)

80. Hutchinson, G.: Circular causal systems in ecology. Annals of the New York Academy of Sciences **50** (Art, 4), 221–46 (1948)

81. Huxel, G.R., McCann, K.: Food web stability: the influence of trophic flows across habitats. Amer. Naturalist **152**(3), 460–469 (1998). https://doi.org/10.1086/286182

82. Ives, A.R., Carpenter, S.R.: Stability and diversity of ecosystems. Science **317**(5834), 58–62 (2007). https://doi.org/10.1126/science.1133258

83. Izhikevich, E.M.: Simple model of spiking neurons. IEEE Trans. Neural Netw. **14**(6), 1569–1572 (2003). https://doi.org/10.1109/TNN.2003.820440

84. Jansen, V.: The dynamics of two diffusively coupled predator-prey populations. Theoret. Popul. Biol. **59**, 119–131 (2001). https://doi.org/10.1006/tpbi.2000.1506

85. Jansen, V., de Ross, A.: The role of space in reducing population cycles. In: Dieckman, U., Law, R., Metz, J. (eds.) The Geometry of Ecological Interactions: Simplifying Spatial Complexity. Cambridge University Press (2000)

86. Jansen, V., Lloyd, A.: Local stability analysis of spatially homogeneous solutions of multi-patch systems. J. Math. Biol. **41**, 232–252 (2000)

87. Jodoin, J.M., Guichard, F.: Non-resource effects of foundation species on meta-ecosystem stability and function. Oikos (2019). https://doi.org/10.1111/oik.06506

88. Jones, C.G., Lawton, J.H., Shachak, M.: Organisms as Ecosystem Engineers. Oikos **69**(3), 373–386 (1994). https://doi.org/10.2307/3545850

89. Kingsland, S.E.: The Evolution of American Ecology, 1890–2000. Johns Hopkins University Press, Baltimore (2005)

90. Klausmeier, C.A., Litchman, E., Daufresne, T., Levin, S.A.: Optimal nitrogen-to-phosphorus stoichiometry of phytoplankton. Nature **429**(6988), 171–174 (2004a). https://doi.org/10.1038/nature02454

91. Klausmeier, C.A., Litchman, E., Levin, S.A.: Phytoplankton growth and stoichiometry under multiple nutrient limitation. Limnol. Oceanogr. **49**(4part2), 1463–1470 (2004b). https://doi.org/10.4319/lo.2004.49.4_part_2.1463

92. Kloosterman, M., Campbell, S., Poulin, F.: A closed NPZ model with delayed nutrient recycling. J. Math. Biol. **68**(4), 815–850 (2014). https://doi.org/10.1007/s00285-013-0646-x

93. Koelle, K., Vandermeer, J.: Dispersal-induced desynchronization: from metapopulations to metacommunities. Ecol. Lett. **8**(2), 167–175 (2005)
94. Krumhansl, K.A., Okamoto, D.K., Rassweiler, A., Novak, M., Bolton, J.J., Cavanaugh, K.C., Connell, S.D., Johnson, C.R., Konar, B., Ling, S.D., Micheli, F., Norderhaug, K.M., Pérez-Matus, A., Sousa-Pinto, I., Reed, D.C., Salomon, A.K., Shears, N.T., Wernberg, T., Anderson, R.J., Barrett, N.S., Buschmann, A.H., Carr, M.H., Caselle, J.E., Derrien-Courtel, S., Edgar, G.J., Edwards, M., Estes, J.A., Goodwin, C., Kenner, M.C., Kushner, D.J., Moy, F.E., Nunn, J., Steneck, R.S., Vásquez, J., Watson, J., Witman, J.D., Byrnes, J.E.K.: Global patterns of kelp forest change over the past half-century. Presented at the (2016). https://doi.org/10.1073/pnas.1606102113
95. Laland, K.N., Odling-Smee, F.J., Feldman, M.W.: Evolutionary consequences of niche construction and their implications for ecology. Presented at the (1999). https://doi.org/10.1073/pnas.96.18.10242
96. Largaespada, C., Guichard, F., Archambault, P.: Meta-ecosystem engineering: nutrient fluxes reveal intraspecific and interspecific feedbacks in fragmented mussel beds. Ecology **93**(2), 324–333 (2012). https://doi.org/10.1890/10-2359.1
97. Lawton, J.H.: Are there general laws in ecology? Oikos **84**(2), 177–192 (1999). https://doi.org/10.2307/3546712
98. Leibold, M., Holyoak, M., Mouquet, N., Amarasekare, P., Chase, J., Hoopes, M., Holt, R., Shurin, J., Law, R., Tilman, D., Loreau, M., Gonzalez, A.: The metacommunity concept: a framework for multi-scale community ecology. Ecol. Lett. **7**(7), 601–613 (2004). https://doi.org/10.1111/j.1461-0248.2004.00608.x
99. Leroux, S., Loreau, M.: Theoretical Perspectives on Bottom-up and Top-down Interactions across Ecosystems, pp. 3–27. Cambridge University Press (2015). https://doi.org/10.1017/CBO9781139924856.002
100. Leroux, S.J., Loreau, M.: Subsidy hypothesis and strength of trophic cascades across ecosystems. Ecol. Lett. **11**(11), 1147–1156 (2008). https://doi.org/10.1111/j.1461-0248.2008.01235.x
101. Leroux, S.J., Loreau, M.: Dynamics of reciprocal pulsed subsidies in local and meta-ecosystems. Ecosystems **15**(1), 48–59 (2012)
102. Leroux, S.J., Wal, E.V., Wiersma, Y.F., Charron, L., Ebel, J.D., Ellis, N.M., Hart, C., Kissler, E., Saunders, P.W., Moudrá, L., Tanner, A.L., Yalcin, S.: Stoichiometric distribution models: Ecological stoichiometry at the landscape extent. Ecol. Lett. **20**(12), 1495–1506 (2017). https://doi.org/10.1111/ele.12859
103. Levins, R.: Qualitative analysis of partially specified systems. Ann. New York Acad. Sci. **231**(Apr22), 123–138 (1974)
104. Liebhold, A., Koenig, W.D., Bjornstad, O.N.: Spatial synchrony in population dynamics. Ann. Rev. Ecol. Evolut. Syst. **35**, 467–490 (2004)
105. Little, C.J., Rizzuto, M., Luhring, T.M., Monk, J.D., Nowicki, R.J., Paseka, R.E., Stegen, J., Symons, C.C., Taub, F.B., Yen, J.: Filling the Information Gap in Meta-Ecosystem Ecology. Tech. rep. (2020). EcoEvoRxiv. https://doi.org/10.32942/osf.io/hc83u
106. Loreau, M.: Material cycling and the stability of ecosystems. Amer. Naturalist **143**(3), 508–513 (1994). https://doi.org/10.2307/2462743
107. Loreau, M.: From Populations to Ecosystems : Theoretical Foundations for a New Ecological Synthesis. Monographs in Population Biology ; 46, Princeton University Press, Princeton (2010)
108. Loreau, M., Hector, A.: Partitioning selection and complementarity in biodiversity experiments. Nature **412**(6842), 72–76 (2001). https://doi.org/10.1038/35083573
109. Loreau, M., Holt, R.D.: Spatial flows and the regulation of ecosystems. Amer. Naturalist **163**(4), 606–615 (2004). https://doi.org/10.1086/382600
110. Loreau, M., Mouquet, N., Holt, R.D.: Meta-ecosystems: a theoretical framework for a spatial ecosystem ecology. Ecol. Lett. **6**(8), 673–679 (2003). https://doi.org/10.1046/j.1461-0248.2003.00483.x
111. Lotka, A.J.: Elements of Physical Biology. Williams & Wilkins (1925)

112. MacArthur, R.R.H.: Fluctuations of animal fluctuations, and a measure of community stability. Ecology **36**(3), 533–536 (1955)
113. Manzoni, S., Trofymow, J.A., Jackson, R.B., Porporato, A.: Stoichiometric controls on carbon, nitrogen, and phosphorus dynamics in decomposing litter. Ecol. Monogr. (2010)
114. Marleau, J.N., Guichard, F.: Meta-ecosystem processes alter ecosystem function and can promote herbivore-mediated coexistence. Ecology **100**(6), e02,699 (2019). https://doi.org/10.1002/ecy.2699
115. Marleau, J.N., Guichard, F., Mallard, F., Loreau, M.: Nutrient flows between ecosystems can destabilize simple food chains. J. Theoret. Biol. **266**(1), 162–174 (2010). https://doi.org/10.1016/j.jtbi.2010.06.022
116. Marleau, J.N., Guichard, F., Loreau, M.: Meta-ecosystem dynamics and functioning on finite spatial networks. Presented at the (2014). https://doi.org/10.1098/rspb.2013.2094
117. Marleau, J.N., Guichard, F., Loreau, M.: Emergence of nutrient co-limitation through movement in stoichiometric meta-ecosystems. Ecol. Lett. **18**(11), 1163–1173 (2015). https://doi.org/10.1111/ele.12495
118. Marleau, J.N., Peller, T., Guichard, F., Gonzalez, A.: Converting ecological currencies: energy, material, and information flows. Trends Ecol. Evolut. (2020). https://doi.org/10.1016/j.tree.2020.07.014
119. Marquet, P.A., Allen, A.P., Brown, J.H., Dunne, J.A., Enquist, B.J., Gillooly, J.F., Gowaty, P.A., Green, J.L., Harte, J., Hubbell, S.P., O'Dwyer, J., Okie, J.G., Ostling, A., Ritchie, M., Storch, D., West, G.B.: On theory in ecology. BioScience **64**(8), 701–710 (2014). https://doi.org/10.1093/biosci/biu098
120. Maser, G.L., Guichard, F., McCann, K.S.: Weak trophic interactions and the balance of enriched metacommunities. J. Theoret. Biol. **247**(2), 337–345 (2007). https://doi.org/10.1016/j.jtbi.2007.03.003
121. Massol, F., Gravel, D., Mouquet, N., Cadotte, M.W., Fukami, T., Leibold, M.A.: Linking community and ecosystem dynamics through spatial ecology. Ecol. Lett. **14**(3), 313–323 (2011). https://doi.org/10.1111/j.1461-0248.2011.01588.x
122. Massol, F., Altermatt, F., Gounand, I., Gravel, D., Leibold, M.A., Mouquet, N.: How life-history traits affect ecosystem properties: effects of dispersal in meta-ecosystems. Oikos **126**(4), 532–546 (2017). https://doi.org/10.1111/oik.03893
123. McCann, K., Hastings, A., Huxel, G.R.: Weak trophic interactions and the balance of nature. Nature **395**(6704), 794–798 (1998). https://doi.org/10.1038/27427
124. McCann, K.S., Rasmussen, J.B., Umbanhowar, J.: The dynamics of spatially coupled food webs. Ecol. Lett. **8**(5), 513–523 (2005). https://doi.org/10.1111/j.1461-0248.2005.00742.x
125. McLeod, A.M., Leroux, S.J.: Incorporating abiotic controls on animal movements in meta-communities. Ecology n/a(n/a):e03, 365 (2021). https://doi.org/10.1002/ecy.3365
126. McLoed, K., Leslie, H.: Ecosystem-Based Management for the Oceans. Island Press, Washington, DC (2009)
127. Menge, B.A., Daley, B.A., Wheeler, P.A., Strub, P.T.: Rocky intertidal oceanography: an association between community structure and nearshore phytoplankton concentration. Limnol. Oceanogr. **42**(1), 57–66 (1997)
128. Menge, B.A., Lubchenco, J., Bracken, M.E.S., Chan, F., Foley, M.M., Freidenburg, T.L., Gaines, S.D., Hudson, G., Krenz, C., Leslie, H., Menge, D.N.L., Russell, R., Webster, M.S.: Coastal oceanography sets the pace of rocky intertidal community dynamics. Presented at the (2003)
129. Assessment, Millennium Ecosystem (ed.): Ecosystems and Human Well-Being: Synthesis. Island Press, Washington, DC (2005)
130. Montbrió, E., Kurths, J., Blasius, B.: Synchronization of two interacting populations of oscillators. Phys. Rev. E Stat. Nonlinear Soft Matter Physics **70**(056), 125 (2004). https://doi.org/10.1103/PhysRevE.70.056125
131. Mouquet, N., Loreau, M.: Coexistence in metacommunities: the regional similarity hypothesis. Amer. Naturalist **159**(4), 420–426 (2002)

132. Mouquet, N., Loreau, M.: Community patterns in source-sink metacommunities. Amer. Naturalist **162**(5), 544–557 (2003). https://doi.org/10.1086/378857

133. Nee, S., May, R.M.: Dynamics of metapopulations: habitat destruction and competitive coexistence. J. Animal Ecol. **61**(1), 37–40 (1992). https://doi.org/10.2307/5506

134. Nicholson, A.J., Bailey, V.A.: The Balance of Animal Populations.—Part I. Proceedings of the Zoological Society of London **105**(3), 551–598 (1935). https://doi.org/10.1111/j.1096-3642.1935.tb01680.x

135. Odling-Smee, F.J., Laland, K.N., Feldman, M.W.: Niche construction. Amer. Naturalist **147**(4), 641–648 (1996). https://doi.org/10.2307/2463239

136. Odum, E.P.: Fundamentals of Ecology. Saunders, Philadelphia (1953)

137. Odum, E.P.: Energy flow in ecosystems: a historical review. Amer. Zoologist **8**(1), 11–18 (1968)

138. Okubo, A., Levin, S.A.: Diffusion and Ecological Problems: Modern Perspectives, 2nd edn. Interdisciplinary Applied Mathematics, Springer, New York (2001). https://doi.org/10.1007/978-1-4757-4978-6

139. Pacala, S.W., Tilman, D.: Limiting similarity in mechanistic and spatial models of plant competition in heterogeneous environments. Amer. Naturalist **143**, 222–257 (1994)

140. Painter, K.J.: Mathematical models for chemotaxis and their applications in self-organisation phenomena. J. Theoret. Biol. **481**, 162–182 (2019). https://doi.org/10.1016/j.jtbi.2018.06.019

141. Peller, T., Andrews, S., Chaves, L., Balbar, A., Leroux, S., Guichard, F.: Integrating ecosystem connectivity into the design of marine protected area networks. Tech. Rep. e26800v1, PeerJ Inc. (2018). https://doi.org/10.7287/peerj.preprints.26800v1

142. Pfister, C.A.: Intertidal invertebrates locally enhance primary production. Ecology **88**(7), 1647–1653 (2007)

143. Pickett, S.T., Kolasa, J., Jones, C.G.: Ecological Understanding?: The Nature of Theory and the Theory of Nature, 2nd edn. Elsevier/Academic Press, Amsterdam (2007)

144. Pillai, P., Gouhier, T.C.: Not even wrong: The spurious measurement of biodiversity's effects on ecosystem functioning. Ecology **100**(7), e02,645 (2019). https://doi.org/10.1002/ecy.2645

145. Polis, G.A., Anderson, W.B., Holt, R.D.: Toward an integration of landscape and food web ecology: the dynamics of spatially subsidized food webs. Ann. Rev. Ecol. Syst. **28**(1), 289–316 (1997). https://doi.org/10.1146/annurev.ecolsys.28.1.289

146. Prins, T., Smaal, A., Dame, R.: A review of the feedbacks between bivalve grazing and ecosystem processes. Aquatic Ecol. **31**(4), 349–359 (1997)

147. Ramiadantsoa, T., Stegner, M.A., Williams, J.W., Ives, A.R.: The potential role of intrinsic processes in generating abrupt and quasi-synchronous tree declines during the Holocene. Ecology **100(2):e02,579**,(2019). https://doi.org/10.1002/ecy.2579

148. Ranta, E., Kaitala, V., Lundberg, P.: The spatial dimension in population fluctuations. Science **278**(5343), 1621–1623 (1997)

149. Reiners, W.A.: Complementary models for ecosystems. Amer. Naturalist **127**(1), 59–73 (1986)

150. Reiners, W.A., Lockwood, J.A.: Philosophical Foundations for the Practices of Ecology. Cambridge University Press, Cambridge, UK (2010)

151. Reuman, D.C., Desharnais, R.A., Costantino, R.F., Ahmad, O.S., Cohen, J.E.: Power spectra reveal the influence of stochasticity on nonlinear population dynamics. Presented at the (2006)

152. Rip, J.M.K., McCann, K.S.: Cross-ecosystem differences in stability and the principle of energy flux. Ecol. Lett. **14**(8), 733–740 (2011). https://doi.org/10.1111/j.1461-0248.2011.01636.x

153. Rohde, K.: Nonequilibrium Ecology. Cambridge (2006)

154. Rosenzweig, M.L.: Paradox of Enrichment: Destabilization of Exploitation Ecosystems in Ecological Time. Science **171**(3969), 385–387 (1971). https://doi.org/10.1126/science.171.3969.385

155. Ruel, J.J., Ayres, M.P.: Jensen's inequality predicts effects of environmental variation. Trends Ecol. Evolut. **14**(9), 361–366 (1999). https://doi.org/10.1016/S0169-5347(99)01664-X

156. Scheiner, S.M., Willig, M.R.: The Theory of Ecology. University of Chicago Press, Chicago, IL (2011)
157. Schiesari, L., Matias, M.G., Prado, P.I., Leibold, M.A., Albert, C.H., Howeth, J.G., Leroux, S.J., Pardini, R., Siqueira, T., Brancalion, P.H.S., Cabeza, M., Coutinho, R.M., Diniz-Filho, J.A.F., Fournier, B., Lahr, D.J.G., Lewinsohn, T.M., Martins, A., Morsello, C., Peres-Neto, P.R., Pillar, V.D., Vázquez, D.P.: Towards an applied metaecology. Perspect. Ecol. Conserv. **17**(4), 172–181 (2019). https://doi.org/10.1016/j.pecon.2019.11.001
158. Schindler, D., Hecky, R., Findlay, D., Stainton, M., Parker, B., Paterson, M., Beaty, K., Lyng, M., Kasian, S.: Eutrophication of lakes cannot be controlled by reducing nitrogen input: Results of a 37-year whole-ecosystem experiment. Presented at the (2008)
159. Schlesinger, W.H., Bernhardt, E.S.: Biogeochemistry, 4th Elsevier, (2020). https://doi.org/10.1016/C2017-0-00311-7
160. Spiecker, B., Gouhier, T.C., Guichard, F.: Reciprocal feedbacks between spatial subsidies and reserve networks in coral reef meta-ecosystems. Ecolog. Appl. **26**(1), 264–278 (2016). https://doi.org/10.1890/15-0478
161. Sterner, R., Elser, J.: Ecological Stoichiometry: The Biology of Elements from Molecules to the Biosphere. Princeton Univ Pr (2002)
162. Strogatz, S.H.: From Kuramoto to Crawford: Exploring the onset of synchronization in populations of coupled oscillators. Physica D: Nonlinear Phenomena **143**(1–4), 1–20 (2000)
163. Takimoto, G., Iwata, T., Murakami, M.: Seasonal subsidy stabilizes food web dynamics: Balance in a heterogeneous landscape: seasonal subsidy and food web stability. Ecol. Res. **17**(4), 433–439 (2002). https://doi.org/10.1046/j.1440-1703.2002.00502.x
164. Tansley, A.G.: The use and abuse of vegetational concepts and terms. Ecology **16**(3), 284–307 (1935)
165. Tilman, D.: Competition and biodiversity in spatially structured habitats. Ecology **75**(1), 2–16 (1994)
166. Tilman, D., Isbell, F., Cowles, J.M.: Biodiversity and ecosystem functioning. Ann. Rev. Ecol. Evolut. Systemat. **45**(1), 471–493 (2014). https://doi.org/10.1146/annurev-ecolsys-120213-091917
167. Tsakalakis, I., Blasius, B., Ryabov, A.: Resource competition and species coexistence in a two-patch metaecosystem model. Theoret. Ecol. **13**(2), 209–221 (2020). https://doi.org/10.1007/s12080-019-00442-w
168. Turchin, P.: Does population ecology have general laws? Oikos **94**(1), 17–26 (2001). https://doi.org/10.1034/j.1600-0706.2001.11310.x
169. Turner, E.L., Bruesewitz, D.A., Mooney, R.F., Montagna, P.A., McClelland, J.W., Sadovski, A., Buskey, E.J.: Comparing performance of five nutrient phytoplankton zooplankton (NPZ) models in coastal lagoons. Ecolog. Model. **277**, 13–26 (2014). https://doi.org/10.1016/j.ecolmodel.2014.01.007
170. Turner, M.G., Gardner, R.H.: Ecosystem Processes in Heterogeneous Landscapes. In: Turner, M.G., Gardner, R.H. (eds.) Landscape Ecology in Theory and Practice: Pattern and Process, pp. 287–332. Springer, New York, NY (2015). https://doi.org/10.1007/978-1-4939-2794-4_8
171. Vandermeer, J.: Oscillating populations and biodiversity maintenance. Bioscience **56**(12), 967–975 (2006)
172. Vitousek, P.M., Porder, S., Houlton, B.Z., Chadwick, O.A.: Terrestrial phosphorus limitation: mechanisms, implications, and nitrogen-phosphorus interactions. Ecolog. Appl. **20**(1), 5–15 (2010). https://doi.org/10.1890/08-0127.1
173. Wang, H., Sterner, R.W., Elser, J.J.: On the "strict homeostasis" assumption in ecological stoichiometry. Ecol. Model. **243**, 81–88 (2012). https://doi.org/10.1016/j.ecolmodel.2012.06.003
174. Wang, S., Loreau, M.: Biodiversity and ecosystem stability across scales in metacommunities. Ecol. Lett. **19**(5), 510–518 (2016). https://doi.org/10.1111/ele.12582
175. Williams, D.W., Liebhold, A.M.: Spatial synchrony of spruce budworm outbreaks in eastern North America. Ecology **81**(10), 2753–2766 (2000)
176. Wootton, J.T.: Local interactions predict large-scale pattern in empirically derived cellular automata. Nature **413**(6858), 841–844 (2001). https://doi.org/10.1038/35101595

Printed in the United States
by Baker & Taylor Publisher Services